Studies in Systems, Decision and Control

Volume 443

The series "Studies in Systems, Decision and Control" (SSDC) covers both new developments and advances, as well as the state of the art, in the various areas of broadly perceived systems, decision making and control–quickly, up to date and with a high quality. The intent is to cover the theory, applications, and perspectives on the state of the art and future developments relevant to systems, decision making, control, complex processes and related areas, as embedded in the fields of engineering, computer science, physics, economics, social and life sciences, as well as the paradigms and methodologies behind them. The series contains monographs, textbooks, lecture notes and edited volumes in systems, decision making and control spanning the areas of Cyber-Physical Systems, Autonomous Systems, Sensor Networks, Control Systems, Energy Systems, Automotive Systems, Biological Systems, Vehicular Networking and Connected Vehicles, Aerospace Systems, Automation, Manufacturing, Smart Grids, Nonlinear Systems, Power Systems, Robotics, Social Systems, Economic Systems and other. Of particular value to both the contributors and the readership are the short publication timeframe and the world-wide distribution and exposure which enable both a wide and rapid dissemination of research output.

Indexed by SCOPUS, DBLP, WTI Frankfurt eG, zbMATH, SCImago.

All books published in the series are submitted for consideration in Web of Science.

Galimkair Mutanov · Askhat Serikbekuly
Editors

Digital Transformation in Sustainable Value Chains and Innovative Infrastructures

Springer

Editors
Galimkair Mutanov
Institute of Information and Computational Technologies
Almaty, Kazakhstan

Askhat Serikbekuly
Up Consulting Llp
Almaty, Kazakhstan

ISSN 2198-4182 ISSN 2198-4190 (electronic)
Studies in Systems, Decision and Control
ISBN 978-3-031-07066-2 ISBN 978-3-031-07067-9 (eBook)
https://doi.org/10.1007/978-3-031-07067-9

This Springer imprint is published by the registered company Springer Nature Switzerland AG
The registered company address is: Gewerbestrasse 11, 6330 Cham, Switzerland

"You have a life, you go with your time." Someone, without opening the door to this bright world, someone in their youth, someone at the peak of their career, and someone in old age left before their descendants ... Since "a person is a guest of life", in this life a person passes a measured time written for him. Sayabek Ziyadin lived a bright, happy, rich but short life. But, he has written hundreds of scientific articles, dozens of monographs and textbooks in the field of economics.
At the source of this book was Sayabek Ziyadin, one of his dreams was to publish the best articles by Kazakhstani authors in Springer. But his dream comes true to his regret after his death.

Sayabek Ziyadin aspired and dreamed to great heights, inherited many scientific works, honored his country, respected younger and older ones, warmed his loved ones, was loyal to his friends, loved all mankind like a brother and became a rising star. Sayabek made a huge contribution to the development of economic science in Kazakhstan and left a huge legacy.
Sayabek is the owner of a big heart, a great scientist, a figure who loves the country and life.
We thank everyone who responded and supported the difficult moments!

Contents

Networks, Infrastructure and Supporting Services in Innovations Development

Digitalization as a Driver for the Development of Strong Regions and Urbanization

Digital Modernization of the System of Public Administration and Law: Prerogatives and Practical Application

Digital Transformation of Education, Science, and Innovations

The Approach to Building a Context-Dependent Sentiment Dictionary

Galimkair Mutanov, Zhanl Mamykova, Vladislav Karyukin, and Sandugash Yessenzhanova

Abstract The rapid development of the Internet has led to a fast increase in the number of news sites and social networks that describe various events in the world and society. Today, authorities consider social networks as a channel of communication with civil society, which allows you to quickly report on the work done and form an actual agenda. Social networks provide an opportunity to regulate the involvement of citizens in the conduct of public affairs, discussion of legal norms, bills, public services, and the activities of public authorities. News posts and comments of a different nature of statements are divided into categories of positive, negative, and neutral assessment. Thus, the problem arises of accurately determining whether these categories of sentiment belong to these categories. Currently, various approaches are implemented to automatically determine the sentiment of texts and sentences: the use of sentiment dictionaries and machine learning algorithms. This article discusses the construction of a more sophisticated sentiment dictionary based on textual data labelled by experts on the content category and sentiment.

Keywords News sites · Social networks · Sentiment analysis · Sentiment dictionary · Machine learning · Supervised learning

G. Mutanov · Z. Mamykova (✉) · V. Karyukin · S. Yessenzhanova
Al-Farabi Kazakh National University, 71, al-Farabi Ave, Almaty 050040, Kazakhstan
e-mail: Zhanl.Mamykova@kaznu.kz

G. Mutanov
e-mail: galimkairmutanov@gmail.com

V. Karyukin
e-mail: Vladislav.Karyukin@kaznu.kz

S. Yessenzhanova
e-mail: esandu92@gmail.com

G. Mutanov and A. Serikbekuly (eds.), *Digital Transformation in Sustainable Value Chains and Innovative Infrastructures*, Studies in Systems, Decision and Control 443,
https://doi.org/10.1007/978-3-031-07067-9_1

1 Introduction

Sentiment analysis is a method designed to automatically determine the sentiment of the vocabulary and opinions of authors on various topics [1]. In this case, the emotional assessment expressed in the text is called the sentiment of the text. The subject of sentiment is the author of the statement; the object of sentiment means what opinion was expressed about, and the sentiment is the emotional attitude of the author.

Sentiment analysis of the text is done with the use of following methods:

1. sentiment dictionary;
2. machine learning algorithms.

In the sentiment dictionary approach, the determination of the sentiment of texts is done with pre-defined sentiment dictionaries [2] that are created manually or automatically. They contain lists of words and related sentiment labels. In most cases, labels are set in the following way:

- positive (1)
- negative (−1)
- neutral (0)

The sentiment of a text is determined by the total number of sentiment values of all words from the created sentiment dictionary that are present is the text. If there is a task of binary text classification as positive or negative, and the words found in the text are predominantly defined as neutral, then labelling texts on categories: politics, society, economics, finance, culture, sports, technology, etc. comes to help. In many cases, a word carries a positive connotation in one category of texts and a negative connotation in another. The opposite picture is also observed.

In machine learning classification algorithms, experts label the text corpus and create documents containing the parameters necessary for the data classification algorithms: text and sentiment [3]. A step of training the classifier consists of the general form of the classification rule with many different parameters. Settings are made on the training set of texts after that.

2 The Review of Existing Analytical Systems

There exists a number of Kazakhstan and foreign monitoring systems of news portals and social networks: iMAS, Alem media monitoring, the OMSystem, Brand Analytics, Microsoft Engagement and others.

iMAS allows you to analyze a public opinion in real time and receive information about critical comments of the society in media, social networks, blogs and forums. Alem media monitoring observes all social networks relevant to Kazakhstan and allows you to keep abreast of various events. Brand Analytics is a Russian system of monitoring user opinions of company brands on social media. Microsoft Social Engagement is a platform for monitoring and interacting with social media space. An opportunity to interact with the public is provided by discussing the products and services of the company.

Developed at Al-Farabi Kazakh National University, the monitoring platform of Kazakhstan media space, the OMSystem (Opinion monitoring system) allows you to monitor the news portals of Kazakhstan and popular social networks: VKontakte, Twitter, Facebook, Instagram and YouTube. The system includes the following features [initiative National Research Work «Development of an information system of monitoring opinions OMS (Opinion monitor system)», 2018, a number of the project's government registration №0118PKИ0331]:

- News articles and comments received via the web crawler;
- Linguistic sentiment dictionary in Russian and Kazakh;
- Calculations of the index of social well-being.

The OMSystem allows you to perform a wide range of functions: automation of routine operations of monitoring the information space, search and processing of a large data array, analysis of opinions of users' perception of events in society, tracking brand mentions, and assessment of the level of social well-being in society.

An important role in the OMSystem is played by the sentiment dictionary, with the help of which the sentiment of texts is determined. To build a sentiment dictionary, many developed lexicons such as Subjectivity Word-list, WordNet-Affect, and SentiWordNet (http://sentiwordnet.isti.cnr.it/) [4] are used. A dictionary list consists of a manually selected set of subjective words, a small unprocessed corpus, and an online dictionary. WordNet-Affect is a lexical resource created on the basis of Princeton WordNet, which contains information about the emotions that words convey. It consists of six main emotions: anger, disgust, fear, joy, sadness, and surprise. Synset is a set of synonyms that represent a meaning or a concept in the dictionary of WordNet (https://www.thefreedictionary.com/synset) [5].

One of the significant problems is the limited resources of the Kazakh language. Many dictionaries are compiled by translation from one language to another (Russian to Kazakh, English to Kazakh).

When we create a sentiment dictionary, the following gradations are most often distinguished:

- binary (positive 1 or negative −1);
- ternary (positive 1, neutral 0, and negative −1);
- multiple (values are set on a scale from −10 to 10). In this case, it is possible to evaluate the sentiment of the sentence/text more sensibly, but more careful work with words is required.

However, when using this dictionary, the sentiment of the text is mainly determined by the presence of words having positive or negative sentiments. Neutral words have little effect on the overall assessment of the text. To expand the capabilities of the sentiment dictionary, it is possible to use categorical features of texts.

3 Methodology of Data Analysis

The following main steps have been used for constructing a context-dependent sentiment dictionary: obtaining data from Internet sources, labelling texts by experts, data preprocessing, classification with machine learning algorithms, and building a context-dependent sentiment dictionary (Fig. 1). The scheme of the OMSystem is also shown in [18].

First, texts from news portals and social networks were obtained using a web crawler [6] and located in the OMSystem database. After that, the texts were labelled by an expert group on thematic categories: politics, society, economics, finance, culture, sports, technology, etc., and sentiment aspects: positive and negative. In total, 2580 articles in Kazakh and Russian were labelled. The division into categories allows you to highlight the thematic focus of the news article and the presence of specific words in it, emphasizing its main idea and content.

The stage of «Text data preprocessing» begins with the removal of elements, symbols, signs, and stop words that do not carry any semantic load. After that, to reduce the number of different words, stemming is performed by removing suffixes and endings from the word so that only the foundation remains, which is the same for all its forms. For Russian words, SnowballStemmer [7–9] from the NLTK library in the Python programming language was used. For words in the Kazakh language, the KazakhStemmer [10] was used. This stemmer removes unnecessary characters and stop words using the regular expression [ˆa-zA-Za-яА-ЯӘІҢҒҮҰҚӨәіңғүұқөһ]. To perform stemming, there is a database of endings and suffixes of the Kazakh language. They are compared to parts of a word, and the removal of the found parts is done in case of coincidence. If a word has a length less than or equal to two, it remains unchanged. This restriction is caused by the fact that some words consisting of two letters are similar to endings and suffixes.

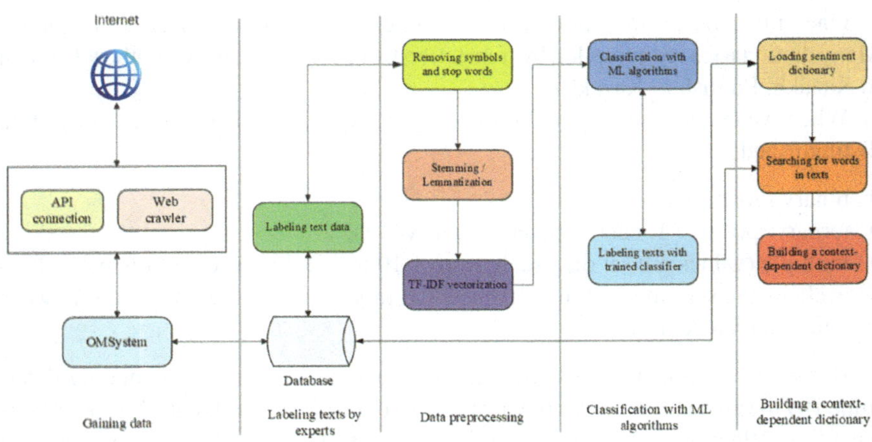

Fig. 1 Stages of building a context-dependent sentiment dictionary

After cleaning the texts, they must be vectorized. An effective vectorization method is the use of the TF-IDF metric [11] for words in each document (one entry in the table is one text document). The calculation of the TF-IDF metric is represented by the formula (1):

$$\mathrm{TF}_{d,w} = \frac{count(w,d)}{count(N,d)} \tag{1}$$

where *count(w,d)* is a number of times a word *w* is met in a document *d*, and *count(N,d)* is a general number of words *N* in a document *d* (2).

$$\mathrm{IDF}_{d,w} = \log \frac{count(D)}{count(w,d)} \tag{2}$$

where *count(D)* is a general number of documents and *count(w,d)* is a number of times a word *w* is met in a document *d*.

The stage «Classification with machine learning algorithms» is implemented after vectorization with the TF-IDF metric, and it consists of the use of the following machine learning algorithms for data classification: Naive Bayes [12, 13], logistic regression, support vector machine, method of k-nearest neighbors [14], a decision tree and a random forest [15].

After the training phase, automatic labelling of new text data is performed. Located in the OMSystem, the sentiment dictionary containing words in the Russian and Kazakh languages is used as a basic dictionary (Fig. 2). All words are divided into positive, negative, and neutral ones. Positive and negative words definitely describe the mood, which is a feature of texts of any category. Neutral words have different aspects depending on the category of texts. For example, news often publishes problems related to public transport. When news with this category appears, such words

	word_id	word	language_id	name	tonal_id	tonal_name
63204	458347	язичник	2	Русский	1	Отрицательный
63205	458348	яица	2	Русский	3	Нейтральный
63206	458349	яицкими	2	Русский	3	Нейтральный
63207	458350	яицо	2	Русский	3	Нейтральный
63208	458351	адамга	1	Казахский	3	Нейтральный
63209	458352	адамгершилиги	1	Казахский	2	Положительный
63210	458354	ак	1	Казахский	3	Нейтральный
63211	458355	ан	1	Казахский	1	Отрицательный

Fig. 2 A part of the sentiment dictionary

as “bus,” “road” and “traffic” can be attributed to the negative sentiment in this particular category with a high degree of probability, but in the general sense these words are neutral.

4 Experimental Results

To evaluate the results of machine learning algorithms: Naive Bayes classifier, logistic regression, support vector machine, k-nearest neighbor method, decision tree and random forest, metrics and graphs of ROC-curves were used. Their calculations are based on true positive (TP), true negative (TN), false positive (FP), and false negative (FN) values.

For the binary classification, metrics are accuracy, precision, recall, and F-measure [16, 17].

- Accuracy is the ratio of the number of correct predictions T to the total number of predictions (3)

$$N = T + F \tag{3}$$

The calculation is carried out according to the formula (4)

$$\text{accuracy} = \frac{T}{N} \tag{4}$$

- Precision is the ratio of objects that really belong to a certain class to all objects assigned to it by the data classification algorithm. The calculation is carried out according to the formula (5)

$$\text{precision} = \frac{TP}{TP + FP} \tag{5}$$

- Recall is the ratio of objects correctly defined by the classifier to all objects of this class in the test sample. The calculation is performed according to the formula (6)

$$\text{recall} = \frac{TP}{TP + FN} \tag{6}$$

- F-measure is a metric that determines the balance of precision and recall. The calculations are performed according to the formula (7)

$$F = 2\frac{\text{precision} * \text{recall}}{\text{precision} + \text{recall}} \tag{7}$$

In the experimental part, we carry out preprocessing of texts in the Russian and Kazakh languages. We balance the classes by sentiment in an approximate ratio of 50% positive and 50% negative to avoid overfitting the model. We divide the data corpus into training at a rate of 70% and testing at 30%. After that, we carry out the training part using each of these machine learning algorithms. To evaluate the results, we test the remaining 30% of the data and evaluate the data classification algorithm using metrics and ROC curves. The ROC curve, or the error curve, is a convenient graphical characteristic of the quality of the binary classifier, showing the dependence of the share of true positive TRP (8) classifications on the share of false positive FPR (9) classifications, where

$$TPR = \frac{TP}{TP + FN} \tag{8}$$

$$FPR = \frac{FP}{FP + TN} \tag{9}$$

The obtained values of the classification metrics for each algorithm are presented in Table 1.

Then we evaluate the quality of data classification algorithms using ROC-curves. The closer the curve is to a value «1», the better the classification result is. Graphics of ROC-curves [16] for classification algorithms are presented in Fig. 3.

In classified texts, we use the sentiment dictionary search. We count a number of positive and negative texts in which a word is present in all categories: politics,

Table 1 Metrics of classification algorithms

Logistic regression	**Value**	**Naïve Bayes classifier**	**Value**
Accuracy	0.87	Accuracy	0.86
Precision	0.88	Precision	0.91
Recall	0.88	Recall	0.81
F-measure	0.88	F-measure	0.86
k-Nearest neighbor method	**Value**	**Support vector machine**	**Value**
Accuracy	0.81	Accuracy	0.86
Precision	0.93	Precision	0.89
Recall	0.69	Recall	0.84
F-measure	0.79	F-measure	0.86
Decision tree	**Value**	**Random forest**	**Value**
Accuracy	0.77	Accuracy	0.815
Precision	0.77	Precision	0.78
Recall	0.81	Recall	0.92
F-measure	0.79	F-measure	0.84

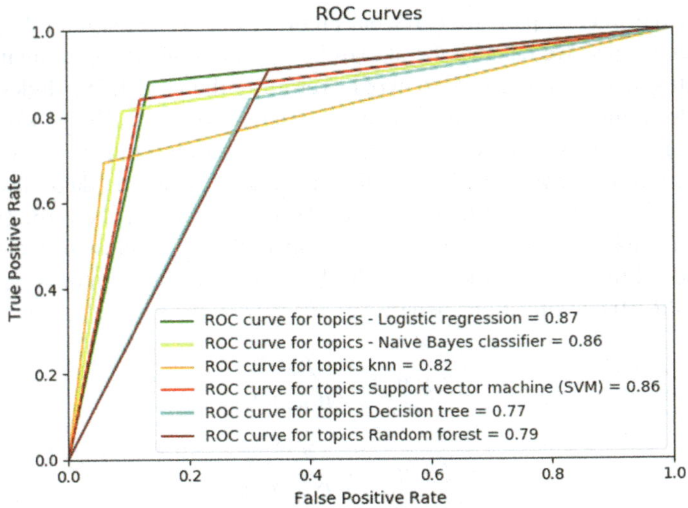

Fig. 3 ROC-curve

society, economics, finance, culture, sports, and technology. Then we remove the words that were not found in the texts. As a result of the search, 7777 Russian words and 4952 Kazakh words were found in the texts.

To determine if a given word in this category is positive or negative, you need to compare the difference between the two meanings. The greater the value of the difference, the more accurately we can say that this word defines the predominant category of the emotional component. From the search results, we do not take into account stop words present in the dictionary since they do not give significant characteristics. From the meaningful words, we compose a new sentiment dictionary in which we write down a word, a category, and a sentiment in this category. An example of the sentiment dictionary is shown in Table 2.

Table 2 Context-dependent sentiment dictionary

Word	Language	Category	Sentiment
доллару	Russian	Economics	Negative
нацвалюты	Russian	Economics	Negative
обменник	Russian	Economics	Negative
процентов	Russian	Finance	Positive
автобуста	Kazakh	Society	Negative
бағдарлама	Kazakh	Society	Negative
		Economics	Positive
жоғары	Kazakh	Society	Negative
		Economics	Positive
жедел жәрдем	Kazakh	Society	Negative

5 Conclusion

The article considers an approach that allows a more accurate assessment of the sentiment of the topics being analyzed using a two-level approach: a dictionary approach and an approach with machine learning algorithms.

The use of both approaches allows building a single flexible sentiment dictionary with dynamic content filling. The developed model of the dictionary will make it possible to label the words by the sentiment and correlate with the categories of the analyzed texts, including neutral words to evaluate in the context of the polarity of the analyzed categories of texts. Such an approach to the construction of the sentiment dictionary defines the «intelligence» of information space monitoring systems.

The application of the developed context-dependent sentiment dictionary in the methodology of the information field will help better understand the mood of society in the most important categories: politics, society, economics, healthcare, etc.

References

1. Mehta, P., Pandya, S.: A review on sentiment analysis methodologies, practices and applications. Int. J. Sci. Technol. Res. **9**(2), 601–609 (2020)
2. Kumar, C.S.P., Babu, L.D.D.: Evolving dictionary based sentiment scoring framework for patient authored text. Evol. Intel. (2020). https://doi.org/10.1007/s12065-020-00366-z
3. Kumar, S. Nezhurina, M.I.: Sentiment analysis on tweets for trains using machine learning. In: 10th International Conference on Soft Computing and Pattern Recognition, SoCPaR 2018; Porto, vol. 942, pp. 94–104. Springer Nature, Switzerland (2020). **DOI:** https://doi.org/10.1007/978-3-030-17065-3_10
4. Baccianella, S., Esuli, A. Sebastiani, F.: SENTIWORDNET 3.0: an enhanced lexical resource for sentiment analysis and opinion mining. In: Proceedings of the 7th International Conference on Language Resources and Evaluation, LREC, pp. 2200–2204 (2010)
5. Fernando, S., Stevenson, M.: Aligning WordNet synsets and Wikipedia articles: AAAI Workshop—Technical Report, vol. WS-10-02, pp. 48–50 (2010)
6. Chaitra P.G., Deepthi V., Vidyashree K.P., Rajini S.: A study on different types of web crawlers. In: Choudhury, S., Mishra, R., Mishra, R., Kumar, A. (eds.) Intelligent Communication, Control and Devices. Advances in Intelligent Systems and Computing, vol. 989. Springer, Singapore (2019). https://doi.org/10.1007/978-981-13-8618-3_80
7. Bounabi, M., El Moutaouakil, K., Satori, K.: A comparison of text classification methods using different stemming techniques. Int. J. Comput. Appl. Technol. (IJCAT) **60**(4) 2019. https://doi.org/10.1504/IJCAT.2019.101171
8. Dolamic, L., Savoy, J.: Indexing and searching strategies for the Russian language. J. Am. Soc. Inform. Sci. Technol. **60**(12), 2540–2547 (2009). https://doi.org/10.1002/asi.21191
9. Fautsch, C., Dolamic, L., Abdou, S., Savoy, L.: Domain-specific IR for German, English and Russian languages. In: 8th Workshop of the Cross-Language Evaluation Forum, CLEF 2007; Budapest; Hungary (2007), Lecture Notes in Computer Science, vol. 5152, pp. 196–199. LNCS (2008). https://doi.org/10.1007/978-3-540-85760-0-26
10. Tukeyev, U., Turganbayeva, A., Abduali, B., Rakhimova, D., Amirova, D., Karibayeva, A.: Lexicon-free stemming for Kazakh language information retrieval. In: IEEE 12th International Conference on Application of Information and Communication Technologies (AICT), Almaty, Kazakhstan, pp. 1–4 (2018). https://doi.org/10.1109/ICAICT.2018.8747021.

11. Salminen, J., Hopf, M., Chowdhury, S.A., et al.: Developing an online hate classifier for multiple social media platforms. Hum. Cent. Comput. Inf. Sci. **10**, 1 (2020). https://doi.org/10.1186/s13673-019-0205-6
12. Wisnu, H., Afif, M., Ruldevyan, Y.: Sentiment analysis on customer satisfaction of digital payment in Indonesia: a comparative study using KNN and Naïve Bayes. In: The 8th Engineering International Conference, Journal of Physics: Conference Series, Semarang; Indonesia, 2019, vol. 1444. IOP Publishing Ltd (2020). https://doi.org/10.1088/1742-6596/1444/1/012034
13. Abd, D.H., Sadiq, A.T., Abbas, A.R.: Political articles categorization based on different Naïve Bayes Models. In: Khalaf, M., Al-Jumeily, D., Lisitsa, A. (eds.) Applied Computing to Support Industry: Innovation and Technology. ACRIT 2019. Communications in Computer and Information Science, vol 1174. Springer, Cham (2020). https://doi.org/10.1007/978-3-030-38752-5_23
14. Sunitha, P.B., Joseph, S., Akhil, P.V.: A study on the performance of supervised algorithms for classification in sentiment analysis. In: TENCON 2019—2019 IEEE Region 10 Conference (TENCON), Kochi, India, pp. 1351–1356 (2019). https://doi.org/10.1109/TENCON.2019.8929530
15. Baheti, R.R., Kinariwala, S.: Detection and analysis of stress using machine learning techniques. Int. J. Eng. Adv. Technol. **9**(1), 335–342 (2019). https://doi.org/10.35940/ijeat.F8573.109119
16. Trivedi, S., Dey, S.: Analysing user sentiment of Indian movie reviews: a probabilistic committee selection model. Electron. Libr. **36**(4), 590–606 (2018). https://doi.org/10.1108/EL-08-2017-0182
17. Wahyudi, M., Kristiyanti D.A: Sentiment analysis of smartphone product review using support vector machine algorithm-based particle swarm optimization. J. Theor. Appl. Inf. Technol. **91**(1), 189–201 (2016)
18. Mutanov, G., Karyukin, V., Mamykova, Z.: Multi-Class Sentiment Analysis of Social Media Data with Machine Learning Algorithms. Comput. Mater. Continua. **69**(1), 913–930 (2021). https://doi.org/10.32604/cmc.2021.017827

The Translation Process in the Context of the Digitalization

Saule Seidenova and Laila Mussaly

Abstract The article considers the problems of translation studies in the digital environment, including the technologization of the translation process and the prospects for its development. Translation studies play an important role in the intellectual, communication, scientific, general cultural development of a country. The current state of translation science in Kazakhstan requires new innovative development and technological renewal. The necessity of digitalization in the translation studies should reveal its technical potential, identify prospective areas and strategies of further development. We use the system, conceptual, aspectual approaches to identify translation problems in the context of digitalization that require a new integrated methodology and solutions within the framework of innovative approaches in education and training in digital translation. The creation of key conditions for the preparation of techniques and personnel for digital translation; improvement of teaching methods and methods, which should ensure the digitalization of translation by competent personnel; focus on the labour market, which should be based on the requirements of the digitalizing translation process; creation of a motivation system for the development of the necessary competencies and the participation of translators in the development of digital policy in Kazakhstan can be solving problems.

Keywords Translation automation · Digital environment · Development paradigm · System analysis · Conceptual method · Classification of information technologies in translation · Innovative potential · Integrative translation studies · Applied aspects of translation

---

S. Seidenova (✉) · L. Mussaly
Al-Farabi Kazakh National University, Al-Farabi Ave. 71, Almaty 050040, Republic of Kazakhstan
e-mail: Saule.Seidenova@kaznu.edu.kz

L. Mussaly
e-mail: Laila.Musaly@kaznu.edu.kz

G. Mutanov and A. Serikbekuly (eds.), *Digital Transformation in Sustainable Value Chains and Innovative Infrastructures*, Studies in Systems, Decision and Control 443,
https://doi.org/10.1007/978-3-031-07067-9_2

1 Introduction

The rapid entry of Kazakhstan into the world community has identified the problem of studying and understanding the urgent problems of studying Kazakhstani translation in the global digital environment. The information space through a global network expands the possibilities of man, allowing him to overcome geographical and political boundaries, making world cultural values accessible to everyone so that they can be contemplated, 'virtualizing' all spheres of human life. The speed of dissemination of information flows leads to a situation of total digitization of social processes and the lives of individuals.

A change in cultural paradigms is taking place right before our eyes in a real-time situation. The term 'informatization' is becoming less and less relevant, the generation that was born and is growing in the process lives in digital format.

2 Literature Review

Nowadays some many investigations are dedicated to problems of translation studies and identity of translators, especially, issues regarding digital literacies in teacher preparation [1, 2].

Moreover, there are some research works about history and key developments in intelligent computer-assisted language learning (ICALL) and the language policy were published in famous journals [3, 4].

Recently translating selves and education services were considered and tested among learners in the best universities around the world. The popular topics are difficulties with indicators of integration and understanding integration as a concept in the multilingual environment [5–8].

Also, there are many problems in the use of machine translation [9, 10].

In them, work with machine translation systems is presented as a step-by-step process that has objective limitations due to technically forced translation by sentences, the need to process vocabulary that is not in the system dictionary, the terminological variability of the source text, the increased length of sentences, the linearity of recognition of collocations.

At the same time, the problem of the formation of the technological culture of the translator is ignored.

3 Methodology

The methodology of the present study combines three approaches to the consideration of the problem of integrativity in translation studies: conceptual, systemic, and aspectual. The conceptual approach involves the preliminary development of the concept of integrative translation as a set of basic theoretical principles on the basis of which translation activities should be carried out, the definition of the components of such activities from integrative positions, that is, taking into account interrelated factors affecting the translation process. This approach provides a unified focus on the description of integrally significant aspects of translation and preservation of continuity in the presentation of translatologically relevant information, including its applicability in translation practice with an emphasis on its quality. The systematic approach makes it possible to take into account the most important aspects of integrativity in translation as much as possible while maintaining the integrity of the aspects of its consideration and their relationship. The aspect approach allows us to justify the choice as the subject of the description of two interconnected faces of the studied translatological problem: identifying the prerequisites for the emergence and development of the main provisions of the integrative translation.

4 Results and Discussion

Recently, a fashionable phrase has appeared as an integrative translation study, which means a scientific direction as a whole, combining the scientific achievements of various sciences regarding the translation process. But in the context of digitalization of the communicative space, translation is of great importance in the social sense.

What is the difference between computerization and digitalization? Informatization is a toolkit in the form of computers, programs, and information cable networks that facilitate a person's work. But this is a tool. The digitalization is a different level. After entering data, the digital system itself solves the tasks, analyzes the information, gives a forecast. This is not artificial intelligence, but a kind of intermediate link between human and artificial intelligence. Computerization, informatization organically enter into a more complex and broader concept of the digitalization.

It is known, information (from the Latin. Informatio - notification, explanation, statement from the Latin. Informare - to give shape) is information (reports, data) about objects of animate or inanimate nature, their properties and mutual influence on each other regardless of form their presentations.

Based on the translation of the word 'technology' (from the Greek 'techne'—art, trade, equipment, we can assume that information technology (IT) is a combination of means and methods of collecting, processing and transmitting information (primary information) in order to obtain information new quality about the state of the object, process or phenomenon (the information product).

As you know, the purpose of information technology is the production of information for human analysis and decision-making on the implementation of certain actions based on it.

In addition to global information technologies, including methods and tools that formalize and allow the use of information resources by society, we use basic information technologies. They are intended for a specific field of application, such as production, research, training. In addition, specific information technologies help to implement data processing when performing functional tasks of users (for example, tasks of inventory, planning, analysis).

As all technologies, information technologies are in constant development and improvement. This is facilitated by the emergence of new equipment, the development of new concepts, methods for organizing data, their transmission, storage and processing, forms of user interaction with technical and other components of the information system.

The hardware and software are technical means of producing information. separately allocated from these funds are software products called 'tools'. For clarity, it is indicated and called 'software tools of information technology.'

The information technology software tools are one or more interconnected software products for a certain type of computer, in which the technology of work allows you to achieve the goal that the user sets.

Regarding the translation process, translation is a set of actions performed by the translator when translating the source (or original) text (ST) into another language. The translation is a means of international communication. The translator makes it possible to exchange information between users of different languages, creating a text in the target language (TL or translation language) that has the same communicative meaning with the original (or original) text (ST).

The translation process includes two mental processes—understanding and verbalization. First, the translator understands the context of ST, that is, reduces the information that he contains to his own mental program, and then he develops this program in TT.

Translation, as one of the activities of mankind, must adapt to the requirements of modern society. The essence of the translation process is constant: the preparation of the draft text is the first stage, and the second is the verification stage, during which the necessary corrections and changes are introduced so that the translation acquires the proper form.

With the advent of new technologies, translation tools and tools are changing. Information technology has created a new electronic culture that has replaced the print culture, and the process of working with documents and information has been greatly simplified and accelerated. Machine translation programs and computer translation tools are a part of this new culture.

In English terminology, information technology in practice divides into two large classes:

1. MT (machine translation). In Russian-language linguistics, this class is known as automatic or machine translation, that is, programs that completely replace the translator with a greater or lesser degree of success.
2. CAT (Computer/automated translation). Unlike MT, CAT systems only automate and facilitate the work of the translator in various aspects. First of all, these are programs that implement the concept of translation memory (translation memory or TM), such as ABBYY Lingvo, Trados, OmegaT, DejaVu, WordFast, etc.

MT made a possible quick translation, but there are still many unsolved problems in this area. Translation relies heavily on context and various connotations of words and phrases. It is not always possible to give a complete formalized context along with the text, therefore automatic translation is limited to specific subjects, and subsequent editing is almost always necessary [9, 10].

As for translation memory (TM), it is a linguistic database that stores source texts and their translations. These texts are divided into segments, which often coincide with sentences. TM makes translation easier—the system 'remembers' the translated phrases (segments) for future use, and the translator does not have to translate the same sentences twice. These programs are very useful for translating texts with a high degree of repetition.

Nowadays a student learning translation business should receive basic knowledge about the fundamental principles and technologies of modern automatic translation systems, so called Machine Translation, MT, and computer-assisted translation (CAT). MT and CAT are the main translation tools that can be used in practice.

This will give real experience of the direct application of CAT-tools in the translation process. Learning to work with translation memory systems as Translation Memory, TM is the core of the course and its main content.

A modern translator must be able to effectively apply automatic translation systems, for example, Prompt and Google Language Tools, and translation memory programms like Trados, OmegaT. The translator must know the operating principles of MT and TM systems (translation memory), as well as other CAT systems (basic terminology, case analysis, concordance, etc.), their place in solving translation problems, be able to work with TheVisualThesaurus.com resource to be able to work with WinZip and WinRar archivers to archive translations and glossaries. To translate a translation to order, a translator must be able to use The Bat, CuteFTP Pro, VentaFax programs, as well as ABBYY FineReader text to work with optical recognition programs.

The importance of translation technologies in various areas of modern life, especially machine and computer translation, is undeniable. The technologies available to translators in the first decade of the 21st century have negative and positive aspects. The main tools used in computer translation are still electronic dictionaries, glossaries, terminology databases, concordances, online bilingual texts and translation memory. In connection with the advent of new technologies, a professional translator aims to acquire new skills in order to remain effective and competitive in his field.

The world has long felt the need for translation technology. As you know, the Internet, with its universal access to information and instant communication between users, created physical and geographical freedom for translators, which was unthinkable in the past. So, the IT department has created a screen culture, which tends to replace the print culture, while printing documents is bypassed, and information is received and transmitted directly through computers (e-mail, databases and other stored information). These computer documents are instantly accessible and can be opened and processed with much more flexibility than printed materials, as a result of which the status of the information itself has changed, becoming temporary or permanent depending on the need. Over the past two decades, we have seen tremendous growth in information technology with the attendant benefits of speed, visual impact, ease of use, convenience and cost-effectiveness. At the same time, with the development of the world market, industry and trade are functioning more than ever on an international scale, with increasing freedom and flexibility in terms of exchanging products and services. The nature and functions of the translation inevitably depend on these changes. Countries need to cooperate in many areas, such as environmental (Greenpeace), economic (free trade agreements) humanitarian (doctors without borders) and educational (exchange programs), etc. Despite the importance of the English language, there is a conviction that that people have the right to use their own language, but the diversity of languages should not be an obstacle to mutual understanding. Solutions to language problems must be found to ensure the free flow of information and facilitate bilateral and multilateral relations [11, 12].

Thus, various aspects of modern life have led to the need for more effective translation methods. Currently, the demand for translations is not satisfied because there are not enough people translators, or because people and organizations do not recognize translation as a complex activity that requires a high level of qualification, and therefore are not ready to pay what it costs. In other words, translation is sometimes avoided because it is considered too expensive. In part, human translation is expensive because human productivity is significantly limited. Statistics vary, but in general, to make a good translation of complex text, the translator cannot process more than 4–6 pages or 2000 words per day. The economic need to find a cheaper solution for international exchange has led to continuous technological progress in terms of translation tools designed to satisfy the translator's need for immediate information availability and inconsistent access to extensive databases.

We need to find out how the study of new technologies (machine translation, electronic dictionaries, terminology databases, bilingual texts, grammatical correspondences and translation memory) affects the relationship between the translator and the texts and what computer tools are really useful for translators, whether new technologies threaten translator's life, does automation mean the disappearance of a translation as we know it.

The first versions of machine translation programs were based on detailed bilingual dictionaries that suggested the number of equivalent words in the target language for each word listed in the source language, as well as a series of rules in word order. The complexity of the task forced developers to continue to improve programs because of the need for a more systematic syntactic orientation. The projects were based on advances in linguistics, especially the development of models of transformational generative grammar, which seemed to open up new possibilities for machine translation [13].

Important events took place in the field of machine translation, when the translation strategy radically changed from grammar-based rules to text-based examples and examples (for example, the Reverse program). Language was no longer perceived as a static entity governed by fixed rules, but as a dynamic body that changes depending on use and users, evolving over time and adapting to social and cultural realities. To this day, machine translation continues to progress. Large companies now use it more, which also increases software sales to the general public. This situation has led to the creation of online machine translation services, such as Altavista, which offer fast email services, web pages, etc. In the right language, as well as the presence of multilingual dictionaries, encyclopedias and free direct access to terminological databases.

At this point, it is important to distinguish between two terms that are closely related and that can confuse laypeople: machine translation (MT) and computer translation (CAT). These two technologies are the result of different approaches. They do not produce the same results and are used in different contexts. MT seeks to collect all the information necessary for translation into one program so that the text can be translated without human intervention. It uses the power of a computer for computation to analyze the structure of an operator or sentence in the source language, break it down into easily translatable elements, and then create an operator with the same structure in the target language. It uses huge multilingual dictionaries, as well as a body of texts that have already been translated. As already mentioned, in the 1980s MT made big promises, but steadily surrenders to computer translation, since the latter responds more realistically to real needs.

CAT uses a number of tools to help the translator work accurately and quickly, the most important of which are terminology databases and translation memory. In fact, the computer offers a new approach to text processing of both source and target text. Working with a digital document gives us inconsistent access to information so that we can use it in accordance with our needs. It becomes easy to analyze sentences of the source text, check the context in which the word or text is used, or, for example, create a list of terms. Similarly, any part of the target text can be changed at any time, and parallel versions can be created for comparison and evaluation. All these aspects have profound implications for the translation, especially from the point of view of evaluating the results, since the translator can work more at ease because of the greater freedom to make changes at any time during the work.

It is important to emphasize that automatic translation systems are not yet capable of producing text that can be used immediately, since languages are highly dependent on context, as well as on various notations and connotations of words and phrases. It is not always possible to provide a complete context within the text itself, so machine translation is limited to specific situations and is considered primarily a means of saving time, and not a replacement for human activity. This requires post-editing in order to get high-quality target text.

Thus, in practice, computer translation is a complex process that includes special tools and technologies that are adaptable to the needs of the translator, who is involved in the whole process, and not just at the editing stage. The computer becomes a workstation where the translator has access to various texts, tools and programs: for example, monolingual and bilingual dictionaries, parallel texts, translated texts in various source and target languages and terminological databases. Each translator can create a personal work environment and transform it in accordance with the needs of a particular task. Thus, computer translation gives the translator flexibility and freedom of movement, as well as immediate access to an amazing range of relevant information. The result is huge time savings.

The following are the most important computer tools at the translator's workplace, from the most basic to the most complex: electronic dictionaries, glossaries and terminology databases.

At first, consultations with electronic or digital dictionaries on a computer are not radically different from using paper dictionaries. However, the benefits will soon become apparent. It takes much less time to type a word on a computer and get an answer than to browse a paper dictionary; There is direct access to related data through links; and you can use several dictionaries at the same time, working with several documents.

The electronic dictionaries are available in several forms: as software that can be installed on a computer; in the form of a CD-ROM and, most importantly, via the Internet. The Google search engine, for example, gives us access to a huge number of monolingual and bilingual dictionaries in many languages, although sometimes it is necessary to become online subscribers, as in the Oxford English Dictionary. Online dictionaries organize material for us from their corps, because they are not just a collection of words in isolation. For example, we can query all words related to one keyword, or all words that come from a particular language. That is, they allow immediate cross-access to information. In addition, the Internet, with its universal access to information and instant communication between users, created physical and geographical freedom for translators, which was unthinkable in the past.

5 Conclusion

Thus, we share the belief that translation has not become a new profession, but that changes will continue and will continue to evolve. Translators must adopt new technologies and learn to use them to the maximum as a means of increasing productivity and improving quality. The final result of the study could be the development of an integrative model for the analysis, assessment and prediction of development trends of modern translation studies in digital format. Ultimately, this will require a new integrated methodology and solutions within the framework of new concepts and innovative approaches. We have a practical interest in the development of a conceptual and methodological framework for the further development of translation studies in the country, as well as for the training of students, undergraduates and doctoral students in translation studies in Kazakhstan. To solve these problems, it is necessary to create key conditions for the preparation of techniques and personnel for digital translation; improving teaching methods and techniques by competent staff; creation of a motivation system for the development of the necessary competencies and the participation of translators in the development of digital policy in Kazakhstan.

References

1. Cronin, M.: Translation and Identity. Routledge, London & New York (2006)
2. Hauck, M., Kurek, M.: Digital literacies in teacher preparation: language, education and technology. In: Steven, L., Thorne, S. (eds.). May Springer International Publishing AG, Springer, Cham, pp. 275–287 (2017)
3. Heift, T.: History and key developments in intelligent computer-assisted language learning (ICALL): language, education and technology. In: Steven, L., Thorne, S. (eds.) May Springer International Publishing AG, Springer, Cham, pp. 289–300 (2017)
4. Spolsky, B.: Language Policy. Cambridge University Press, Cambridge (2004)
5. Tipton, R.: Translating/ed selves and voices: language support provisions for victims of domestic violence in a British third sector organization. Transl. Interpret. Stud. **13**(2), 163–184 (2018)
6. Vasey, H., et al.: ESOL at Manchester Adult Education Service (MAES): Experiences of Entry 3 and Level 1 Learners, Multilingual Manchester in Conjunction with Manchester Adult Education Service (Maes). University of Manchester (2018)
7. Ager, A., Strang, A.B.: Indicators of Integration: Final Report. London, HMSO (2004)
8. Ager, A., Strang, A.B.: Understanding integration: a conceptual framework. J. Refug. Stud. **21**(2), 166–191 (2008)
9. Blommaert, J., Collins, J., Slembrouck, S.: Spaces of multilingualism. Language and Community **25**(3), 197–216 (2005)
10. Garbovskiy, N.K.: System model of translation studies. trans-disciplinarity and the system of scientific knowledge. Mosc. Univ. Bull., Transl. Theory **1**, 3–19 (2015)
11. Garbovskiy, N.K.: Theory of Translation [Translation Theory]. Publishing of Moscow university, Moscow (2004)
12. Polikarpov, A.M.: Integrative model of Translation process. Linguist. Transl.: Collect. Sci. Pap. Mater. **1**, 168–179 (2011)
13. Polikarpov, A.M.: Integrative Translation Studies and Philosophical hermeneutics. G Genres and types of texts in scientific and media discourse. Int. Collect. Sci. Pap. **12**, 280–295 (2014)

The Development of Structure Health Based on Digital Technologies

Bibinur Sydykova, Aliya Atabayeva, and Kundyzay Tobzhanova

Abstract The article provides examples in the field of healthcare, medical development, and what positive aspects digital technologies have in the field of healthcare. Digitalization in healthcare offers the ability to continuously monitor, change behavior, and personalize interventions at low cost, potentially easing the burden of chronic diseases in cost-constrained healthcare systems. Healthcare is an industry that requires large amounts of data. Methods for collecting, storing, and retrieving clinical data have evolved within a global electronic ecosystem that includes computers, mobile devices, the Internet, social networks, and a more interconnected world. Image research, laboratory analysis, drug delivery systems, diagnostic devices, robotic surgery, computer prosthetics, and bio-monitors all rely on computers and electronic data systems.

Keywords Healthcare · Digital technologies · Digitalization · u-Healthcare · Medical development · system medicine

1 Introduction

As we enter the health information age, digital health technologies offer significant opportunities to optimize both health care delivery and clinical research. Despite their potential, the use of such information technologies in clinical care and research faces serious data quality, privacy, and regulatory challenges [1].

The Internet and its associated digital technologies have generated a rapid growth of the information revolution in healthcare. Emerging evidence supports the positive impact of interactive e-health interactive programs, although many challenges

B. Sydykova (✉) · A. Atabayeva
Semey Medical University, Semey, Kazakhstan
e-mail: bibinursk@mail.ru

K. Tobzhanova
Al-Farabi Kazakh National University, al-Farabi Ave. 71, Almaty 050040, Republic of Kazakhstan

G. Mutanov and A. Serikbekuly (eds.), *Digital Transformation in Sustainable Value Chains and Innovative Infrastructures*, Studies in Systems, Decision and Control 443,
https://doi.org/10.1007/978-3-031-07067-9_3

remain regarding research approaches to methodology, implementation, and evaluation. The ubiquitous network, the interweaving of computers, consumer electronics, automobiles, home appliances, medical equipment, transportation systems, animals and plants through the overlap of electronic spaces and physical spaces, is expected to provide an environment that improves our lives by increasing convenience, efficiency and security in all areas [2].

The modern healthcare industry emphasizes safety, efficiency, a patient-centered approach, timeliness, and balance. u-healthcare allows you to safely provide relevant services from anywhere at any time. To explain the emergence of u-healthcare, it is necessary to refer first to the progress in information technology and medical technology, and then to the desire and readiness of medical institutions to adopt a new concept of service and the increased demand for medical care. The digitalization of information, the introduction of broadband, and the leaps made in medical technology in recent years have provided the technological potential needed to achieve u-healthcare [3].

Over the past 10 years, many countries have adopted policies that encourage health systems to implement an extremely complex set of interconnected, often externally developed, software applications that together create an electronic health record (EHR) [1]. Using EHRs involves a wide range of stakeholders, often with conflicting priorities and requirements. An example of such conflicts is the following scenario. Most doctors would ideally like to dictate, write, or print a brief, highly technical description of a patient's problem and treatment plans for their progress notes and orders, and then have someone else responsible for entering the information in the encoded fields needed to support real-time clinical decisions, quality reporting, and billing [2]. Patients want an easy-to-understand explanation of their main medical problems and a clear explanation of what they should do [4].

2 Literature Review

There is growing interest in electronic access to medical information and the use of digital data, both for disease tracking and health tracking. The widespread use of health information technology (IT) has the potential to increase patient's access to their health information and contribute to future goals for the development of patient-centered care. For example, tools such as personal medical records, patient portals, and various mobile health apps (mHealth) have been developed to help patients participate in their own care. Already, a significant number of patients use health IT; therefore, it is very important that the patient's health is adapted to their needs [5].

Patients can control their own physiological parameters and medical events using mobile apps. The problem is how to attract patients to regular use. The key issue seems to be convincing the patient that the doctor will review the data sent to the patient's medical records service (PHR) in a timely manner. On the other hand, we must provide doctors with convenient access to such data and not overload it. The dynamics of clinical indicators of diabetes, hypertension, and anticoagulants are vital for successful treatment, so these patients and their doctors are promising for innovative methods. A specialized medical messenger allows the patient to ask questions to the doctor at the moment when they arise, without disturbing the doctor's life using his personal mobile phone or email. We expect this messaging to serve as the basis for an innovative interactive medical history, managed not only by the doctor, but also by the patient. Personal monitoring is also used to assess "treatment adherence": a system of reminders about medication intake and measurements, the results of which are made available to the doctor in accordance with their chosen intelligent algorithms [6].

3 Methodology

The following research methods were used in the preparation of this research paper:

General scientific method: legislation and world experience of transition to a digital environment, implementation of digitalization in modern society.

Comparative method: analogies of digitalization were carried out. The main components of digital healthcare are the following:

1. Electronic document flow between a doctor, patient, and medical organization;
2. Application of telemedicine technologies in the provision of medical care;
3. Application of mathematical methods (including methods of artificial intelligence, big data processing) in the processing of medical data.

4 Results

Systems biology and the digital revolution are together transforming healthcare into a proactive P4 drug that is predictive, preventative, personalized, and collaborative. Systems biology—a holistic, global and integrated approach-gave birth to systems medicine, a systematic approach to health and disease. System medicine promises:

- provide a deep understanding of the mechanisms of the disease;
- make blood a diagnostic window for monitoring health and disease for an individual;
- to divide a complex disease on their separate subtypes for impedancometry with appropriate drugs;
- provide new approaches to drug target detection and create indicators to assess well-being.

P4 medicine, the clinical face of systems medicine, has two main goals: to quantify health and demystify disease. Patients and consumers will be the main driving force in the implementation of P4 medicines through their participation in medical-focused networks aimed at improving their own healthcare. P4 medicine has startling implications for society—including the ability to turn around ever-increasing health care costs. The task of communicating P4 medicine to patients and consumers consists of two parts: first, to invent strategies and technologies that will allow the use of P4 medicine, and second, to consider the impact of P4 drugs on society, including the main ethical, social, legal, regulatory, and economic issues. Managing social problems will create the most serious problems. Strategic partnerships of various types will be needed to bring the P4 drug to patients [7].

Personalized medicine promises to predict, prevent, and treat diseases that are tailored to people's needs. New technologies for detailed biological profiling of people at the molecular level were crucial at the beginning of the transition to personalized medicine; further new technologies will be needed if the dream becomes a reality. Standardization, integration and harmonization are key factors in the development of personalized medicine technologies. For example, tools and processes for data collection and analysis should be standardized across all research sites. Research activities at different sites should be integrated to maximize synergies, and research should be integrated with health care to ensure effective translation. There should also be harmonization between scientific practices in various research centers, between science and health, and between science, health and society as a whole, including ethical and regulatory frameworks, prevailing political and cultural ethnos, and patient/citizen expectations.

Many errors that occur in the health sector are related to the lack of important medical information about the patient. According to researchers, the use of information and communication technologies (ICTs) promises to increase the availability of medical information, which is extremely important for patient safety. In addition, new developments in information and computing technologies will lead to dramatic changes in the healthcare environment. An extensive and multi-level infrastructure of ubiquitous computing technologies and applications is emerging. Currently, the widespread use of mobile phones, laptops, Wi-Fi, Bluetooth, personal digital assistants (PDAs) and various types of sensor devices based on digital and radio frequency identification (RFID) technologies is penetrating the healthcare environment [8].

Creative use of new mobile and medical information and sensor technologies (mHealth) can reduce the cost of medical care and improve well-being in many ways. These applications are being developed in various fields, but thorough research is needed to explore the potential as well as challenges of using mobile technology to improve health outcomes. Currently, there is little evidence of the effectiveness of mobile healthcare. While these technologies may be attractive and seemingly harmless, research is needed to assess when, where, and for whom mHealth devices, applications, and systems are effective.

To outline an approach to collecting evidence in the field of mobile health that would ensure that research is conducted on a rigorous empirical and theoretical basis, on August 16, 2011, researchers gathered for a workshop on mobile health evidence in NIH [9].

Currently, more than 165,000 mobile health apps are available in major app stores, the vast majority of which are intended for patients. The top two categories are health and Wellness management apps, while other categories include self-diagnosis, medication reminder, and electronic patient portal apps. Applications specific to physical medicine and rehabilitation are also considered. These applications can provide low-cost, round-the-clock access to high-quality, evidence-based health information for end users globally. However, they have yet to live up to their potential due to numerous barriers, including a lack of regulatory oversight, limited evidence-based literature, and privacy and security concerns. Future directions may include improving data integration into the healthcare system, a feature-compatible application platform that provides access to electronic medical card data, cloud-based personal medical records across the healthcare network and increasing prescription applications by healthcare providers. For consumer mobile health apps to fully contribute to improving the quality of care and treatment of chronic diseases, all stakeholders in the ecosystem must work together to overcome significant barriers [10].

Mobile apps can become the main source of health recommendations; however, problems with their quality and safety remain. Governments, clinicians, and health researchers are trying to determine how to adequately evaluate the content and features of mobile health apps to guide consumers to apps that will effectively and safely support their health. Researchers adapt existing research methods, for example, for systematic reviews, and explore new methods to achieve this goal. The purpose of this systematic review was to describe common and emerging methods of search, data extraction, and analysis in the context of mobile medical applications and commercial app stores [11] (Table 1).

The movement towards the development of digital medicine is formed by rapidly developing IT technologies. It systems must be implemented. For example, today in America and a number of other countries, IT is used to support medical decisions, information exchange with other medical institutions, with insurance companies, electronic prescription, obtaining laboratory data and using medical information systems (MIS) during admission (Table 2).

5 Conclusion

Telecommunications, multimedia and computer technologies will make significant changes in the management of health care. New methods of presenting patient's medical records using computer technology products and services allow unlimited information exchange. Training using multimedia methods and the Internet is available everywhere and at any time, like surgical modeling, robotics and virtual reality. Thanks to computer and it technologies, surgeons will be able to acquire, evaluate and validate new surgical procedures or concepts from any geographical location. Virtual reality is a major improvement in medical image processing.

Ultimately, the Internet offers seamless communication channels between health care providers.Information and digitalization technologies will undoubtedly contribute to health care worldwide.

Table 1 PEST-analysis

Political factors	**Economic forces**
Changes to tax laws, consumer protection and employment rules, and insurance mandates are all elements in the political sphere that could have an impact on health care. For example, a change in tax policy may require an adjustment to a strategy that either takes advantage of increased government spending on health care or reduces government subsidies. Or changes to labor laws, such as the 2016 legislation that affected overtime requirements, could mean major changes in staffing and overtime requirements	Unemployment, inflation, and interest rates are examples of economic problems that directly or indirectly affect the financial performance of health organizations. These changing conditions may affect government spending policies and your purchasing power. For example, if you manufacture medical equipment, a high unemployment rate will mean that fewer people will be able to buy your products. Or, more people losing their jobs means more loss of health insurance coverage, which will affect the types of health services that people are looking for
Social factors	**Technological factor**
The practice of environmental sustainability has both environmental and financial advantages for health organizations. Many hospitals are currently seeking LEED certification for green building design; waste management programs are also under scrutiny. Leadership in green environment practices also illustrates corporate social responsibility (CSR), which should be a key part of any strategic plan	Technological advances that are specific to hospitals and healthcare providers may have different effects on your overall performance. The equipment produced and used is constantly changing, as well as the development of new treatment methods. And new technologies outside of healthcare can also affect how an organization communicates with its consumers, as in app development or digital marketing. Planning ahead to take advantage of these achievements can mean good opportunities for growth
Environmental factors	**Legal factors**
The practice of environmental sustainability has both environmental and financial advantages for health organizations. Many hospitals are currently seeking LEED certification for green building design; waste management programs are also under scrutiny. Leadership in green environment practices also illustrates corporate social responsibility (CSR), which should be a key part of any strategic plan	Failure to comply with new laws and legal procedures governing health care can lead to an immediate failure of your strategy. The legal implications of pending litigation, HIPAA compliance, potential hospital mergers, and more should be taken into account as part of your overall strategy

Table 2 SWOT-analysis

Strenght	Weaknesses
• effective control of healthcare expenses • standardization of clinical practice accelerating the pace of healthcare digitalization; • use of the most modern medical equipment • focus on improving health care	• lack of funding and resources to support programs • fluctuations in continuity of care due to poor communication, • insufficient training of management personnel • poor use of medical Informatics and lack of financial resources • low level of literacy and public awareness of harm • imbalance of personnel in the system of health care;
Opportunities	**Threats**
• development of new health programs • an increase in funds for improvement of medical Informatics • implementation of artificial intelligence and full digitalization; • mobile and web applications for information and engagement • the population's contribution to protecting their own health; • the development of personalized medicine; • the development of medical tourism; • introduction of innovative treatment methods and technologies through Startup	• economic or political instability • changes in policies and legislation that lead to budget deficits • increased demand for expensive medical technologies • the budget deficit at the state level and the Federal level, and increased pressures to reduce health care costs • outflow of highly qualified personnel; • global climate change, air, soil and water pollution; • global outbreaks of highly dangerous infections and their rapid spread • distribution via air service • the burden of non-communicable diseases; • disinformation of the population through unreliable sources; • refusal of vaccination by the population, reduction of patient compliance

References

1. Sharma, A., Harrington, R.A., McClellan, M.B., Turakhia, M.P., Eapen, Z.J., Steinhubl, S., Green, E.M.: Using digital health technology to better generate evidence and deliver evidence-based care. J. Am. Coll. Cardiol. **71**(23), 2680–2690 (2018)
2. Ahern, D.K.: Challenges and opportunities of eHealth research. Am. J. Prev. Med. **32**(5), 75–82 (2007)
3. Baldwin, J.L., Singh, H., Sittig, D.F., Giardina, T.D.: Patient portals and health apps: Pitfalls, promises, and what one might learn from the other. Healthcare **5**(3), 81–85 (2017) (Elsevier)
4. Sittig, D.F., Belmont, E., Singh, H.: Improving the safety of health information technology requires shared responsibility: it is time weal step up. Healthcare **6**(1), 7–12 (2018) (Elsevier)
5. Nundy, S., Oswald, J.: Relationship-centered care: a new paradigm for population health management. Healthcare **2**(4), 216–219 (2014) (Elsevier)
6. Shklovsky-Kordi, N.: Engaging patients, empowering doctors in digitalization of healthcare //International Conference on Health Information Science, pp. 40–44. Springer, Cham (2017)

7. Hood, L., Flores, M.: A personal view on systems medicine and the emergence of proactive P4 medicine: predictive, preventive, personalized and participatory. New Biotechnol. **29**(6), 613–624 (2012) (Elsevier)
8. Mutter, D., Bouras, G., Marescaux, J.: Digital technologies and quality improvement in cancer surgery. Eur. J. Surg. Oncol. (EJSO) **31**(6), 689–694 (2005)
9. Kumar, S., Nilsen, W.J., Abernethy, A., Atienza, A., Patrick, K., Pavel, M., Hedeker, D.: Mobile health technology evaluation: the mHealth evidence workshop. Am. J. Prev. Med. **45**(2), 228–236 (2013)
10. Kao, C.K., Liebovitz, D.M.: Consumer mobile health apps: current state, barriers, and future directions. PM&R **9**(5), 106–115 (2017)
11. Grundy, Q.H., Wang, Z., Bero, L.A.: Challenges in assessing mobile health app quality: a systematic review of prevalent and innovative methods. Am. J. Prev. Med. **51**(6), 1051–1059 (2016)

Application of ServQual as a Quality Management Technique in Higher Education: The Case of Kazakh National University

Gulden Manarbek and Saltanat Kondybayeva

Abstract The purpose of this paper is to study the applicability of a measurement tool ServQual, in defining the quality of provided service in higher education from perspectives of student population. The paper firstly reviews the available literature on ServQual implementation in higher education. Secondly, the paper attempts to investigate validity and reliability of the SERVQual model in assessing the quality of education by surveying students of al-Farabi Kazakh National University in Kazakhstan. A questionnaire study developed following the well-known business model SERVQual. A total of 322 undergraduates from the national university participated in this survey. The results of the research outlined the higher the expectations are, the bigger the gap is. The findings indicated that expectations did not comply with the administrative services of the university when the mean value of expectation was '5' and '4.5'. However, when the mean score was '4' and '3.5', there was significant value in the obtained results. Findings of this research could assist higher education managers to improve the quality of administrative services, meeting students' expectations. This study adds value to the Kazakhstani scholarly literature by providing research on the applicability of the SERVQual model in Kazakhstani higher education.

Keywords Service quality · Quality management · Quality management tool · ServQual method · Higher education

---

G. Manarbek (✉)
Higher School of Economics and Business, Department of Management, Al-Farabi Kazakh National University, al-Farabi Ave. 71, Almaty 050040, Republic of Kazakhstan
e-mail: Gulden.Manarbek@kaznu.edu.kz

S. Kondybayeva
Higher School of Economics and Business, Department of Economics, Al-Farabi Kazakh National University, Almaty, Kazakhstan

G. Mutanov and A. Serikbekuly (eds.), *Digital Transformation in Sustainable Value Chains and Innovative Infrastructures*, Studies in Systems, Decision and Control 443,
https://doi.org/10.1007/978-3-031-07067-9_4

1 Introduction

High competitive higher education environment has led university managers to seek for innovative approaches and tools for quality management of education. Quality education in turn is a guarantee of new students, foreign students, high reputation at the labour market, as well as a key for maintaining university position at the local and international educational area.

Currently, the national and international competitive world has challenged higher education institutions to apply business models of service quality and other marketing activities to manage effectively the quality of education. In this regard, the representatives of the higher education system have already acknowledged the vital importance of service quality and its serious impact on quality management of higher education. Application of various business models and quality tools in higher education is becoming more popular among university managers. In this paper, the research focuses on application of the well-known quality technique ServQual in the context of Kazakhstani higher education, where expectations are outlined by already outlined variables.

Obviously, òne of the factors of quality improvement in higher education is the measurement of satisfaction of external stakeholders. In this regard, University authorities have realized the necessity of applying quality tools of business field to improve the quality of provided services. In the same manner, in recent times higher education institutions in Kazakhstan are being exposed to national and international competitions in the wake of new changes in the higher education system, as well as challenges in the labour market for the leading position. Today, Kazakhstani universities are demonstrating their high-quality performance through participation in world university rankings, improving quality of educational programmes through international accreditation and modernization of degree programmes, as well as introducing innovative study programmes to attract more smart students. In the same manner, university management regularly carries out quality assessment procedures to reveal weak and strong sides of provided educational services. However, there is almost lack of theoretical and empirical studies on application of the ServQual method in the context of KZ HE. In this regard, this paper individually has researched satisfaction level of students with perceived services in accordance with expectations using the ServQual method.

Following, the present research attempts to identify applicability of the SERVQual method in diagnosing students' satisfaction level with provided services by the university management.

2 Literature Review

In the global literature, there is a growing interest to the discussion of the practical application of the ServQual model in the field of HE. The research conducted about the service quality of the business sector is tremendous. With the expansion of the HEIs role in the society and economy, there is a growing interest in studying service quality in quality management of higher education.

The current research mainly concentrates on identifying and measuring main quality tools for effective management of quality in higher education. Recently, the concept of "service quality" in the higher education sector has attracted significant attention of academies, managers of higher education institutions and quality experts. Besides, despite development of some fundamental measurement tools of service quality from perspectives of students, the principles or criteria towards quality management requires thorough studies. In the literature, there is no unique definition of "service quality". Some believe that it is the elimination of the gap between expectations and perceptions of customers with provided service. Other scholars have discussed the relationship between service quality and student satisfaction as well as a crucial role of student satisfaction research in innovative management of higher education [1]. According to Lewis and Smith [2], quality of service in (service) industry is a key trigger to competition.

The definition given to the concept "service quality" in the business sector is abundant and the interest in service quality studies grew in the early 1980s. For instance, Lewis and Booms [3] claimed that service quality is assessed when the level of customers' satisfaction riches their expectations. As for quality measurement tools, Firdaus Abdullah believes that a new measurement instrument HEdPERF (Higher Education PERFormance) of service quality in tertiary education consisting of a six-factor structure, could be a useful tool to enhance the quality of educational services and to manage quality effectively in light of global competition at labour market [4, 5]. He insists on university managers to assess all six-dimensions (non-academic aspects, academic aspects, reputation, access, programme issues and understanding) to identify the main strengths and weaknesses of the university activity for improvement. In this regard, there is a great challenge for university managers to outline key quality dimensions, that require greater attention than others, so allocation of financial, material and human resources for the improvement of quality education will be effective.

In recent years, foreign literature has mainly focused on studying the quality of provided services by higher education institutions. With the rise of competitiveness for quality graduates and for sustaining position at educational and labour markets, the role of quality education has become a crucial issue for university management. In this regard, plenty of articles, books have been dedicated to the implementation of business-models in higher education to define the quality of educational services.

Today, scholars discuss the effectiveness of implementing a measurement tool developed by Parasuraman [6] ServQual scale, which is popular in the business sector for determining the quality of service and satisfaction level from customers' perspectives and expectations.

Although it may be true that application of business quality measurement scales is an effective tool for assessing customer's satisfaction, there is a school of thought that, collected viewpoints of students regarding the quality of provided services, do not make any sense in regards to university quality management policies and processes [7]. Other scholars, as Cloutier and Richards believe that, 'measuring students' satisfaction in the sphere of higher education is a great ordeal for quality managers of HEIs [8].

The number of researchers argues that quality of education is a major backbone of higher education institution [9] as well as the quality of provided service in higher education is a prerequisite of university success [10]. As far as "quality education' is concerned, Marrs et al. ([11], p. 965 in Soni S., Govender K., 2018) claim that branding is one of the most significant strategic marketing approaches for quality assurance in light of the economic and financial challenges [12]. However, there is another group of scholars, who investigated the impact of service quality in HE and its contribution to the performance of HEIs [13]. Soares et al. [14] explored the application of the ServQual model in Brazilian public institutions to assess quality of administrative services and identified 'tangibility' dimension to be the one with positive outcomes among five. As for Cristina et al. [15], they carried out an empirical study identifying which dimension serves best to define quality of educational services from perspectives of student population and differences in perceived quality between private and public HEIs using the ServQual Model.

The literature review provided between 1991 and 2018 from the Web of Science Core Collection database has shown that there are numerous studies where the application of the SERVQUAL instrument as a measurement tool to assess service quality in higher education is popular Table 1.

The findings of the literature review in the well-known and well-respected database, as well as according to Robinson [16] and Waugh [17] 'the most enduring, popular and the best researched method of evaluating service quality is the ServQual Model', developed by Parasuraman et al. [6, 18]. Brysland and Curry [19] claimed

Table 1 Results of qualitative literature analysis within the web of science categories "Management" "Business", "Economics", "Business Economics", "Education Educational Research"

Model	Total number/categories	Citation in total/in higher education	Number in higher education
European Foundation for Quality Management (EFQM)	32/15	84/8	4
Balanced Scorecard (BSC)	1062/549	12,587/71	19
Malcolm Baldridge National Quality Award (MBNQA)	9/8	17/–	–
The ISO 9000 standards	1402/196	2686/11	4
Business Process Re-engineering (BPR)	144/72	620/–	–
SERVQUAL model	**248/138**	**12,541/152**	**19**
Six Sigma	1907/730	7499/113	22

Source Authors' own reproduction

that the SERVQual model is the validated instrument, which could be applied as a tool for benchmarking practices.

There is a number of empirical studies in global literature outlining the SERVQUAL model as an effective measuring service quality tool in the HE environment [20–22].

However, findings of the literature research obtained in the Web of Science core collection database, show that there is a lack of studies conducted on service quality of higher education in case of Kazakhstan [23]. Only the single document found in the Web of Science CC is related to the evaluation of the educational services based on Mamdani's fuzzy inference algorithm [24].

After gaining the independence from the Soviet Union, the government of Kazakhstan has allocated all resources to quality assurance of higher education, acknowledging its role in the prosperity of the nation and improvement of economy. As an evidence, several reforms have been introduced—the structure of higher education system transferred from being completely government-controlled and publicly funded to a complex system with partially government controlled structure, performance-based funding and granting autonomous to HEIs to be competitive in a national and international market [25]. In this regard, with the shift of university governance of national universities in Kazakhstan to market-oriented one, growing autonomy and competition at educational and labour markets have made Kazakhstani university administrators to act as business actors (http://adilet.zan.kz/rus/docs/Z1500000398#z6). The current reforms and changes being introduced in HE system of KZ are shaping and transforming quality management techniques and approaches. Mainly, the nature of universities are becoming more market-oriented with the growing academic, financial and managerial autonomy. In this context, HEIs are being associated as service industry, aimed to satisfy demands of external stakeholders. Thus, the current research paper as a pioneer attempts to apply the well-known service quality tool in KZ HE and investigates its reliability as a pilot approach at the regional level.

3 Methodology

ServQual method has a broad scope of applications in different sectors of the literature. In this study, the quality of service from the perspectives of students is considered and defined. The expectation means have been structured based on foreign practice on application of ServQual questionnaire, which includes 22 items within 5 dimensions. The survey was distributed to students of 3–4 years study; respondents were to answer only in the framework of their perceptions, since the most common expectation variables have been identified as ideal ones. As a result, the research found out that the higher the expectations are, the bigger the gap is.

The administration body of the university was informed about the purpose of the study. The first section of the developed questionnaire included the demographic questions and information about students' faculty, year as well as their major of study.

The basic part of the survey, defined for outlining the level of satisfaction among students, consisted of 22 specific questions, which encompassed five dimensions developed by Parasuraman et al. [6]. Parasuraman et al. [6] proposed a measurement scale "ServQual" that uses the following dimensions: "tangibles"—refers to physical facilities, material-technical base; "reliability"—refers to the ability of managers to provide services dependably and accurately; "responsiveness"—willingness to ensure quick and productive service; "assurance'—refers to ability of employees to ensure confidence to customers; "empathy"—readiness to provide each customer with personal service [26].

For the main part of the questionnaire, a 5 point Likert scale from 5 (strongly agree) to 1 (2—strongly disagree, 1—do not know) was adapted. Before the implementation of the questionnaire, it has been checked, revised several times.

The focus of the research concentrated on 3 and 4 years students as the basic criterion for selection of the samples since they could provide more accurate and objective answer having more experience studying at the target university. In the same manner, we attempted to encompass all faculties and to survey the equal number of students at each school to acquire more precise and empirical results. It is worth to note that an additional category for the ethnicity of respondents has been provided, as Kazakhstan is a multi-national country.

We carried out the research in the case of only al-Farabi national university, since it is the leading and the largest university in Kazakhstan. The reason for selecting the particular national university stems from the leading position of the university in the country, and the university being the oldest one in the region. In the same manner, in 2020, al-Farabi Kazakh National University has positioned itself at 165 rank in the QS World University Rankings 2020 and the British Agency "Times Higher Education" has ranked the national university as the best university in Central Asia. This in turn demonstrates high performance and high competitiveness of university at global education and science space. Since it is the first and pilot study about the application of the business quality tool ServQual in higher education at the regional level, we approached to the study of its applicability based on a national university, which enroll the majority of student population every year in the country.

In total 322 respondents participated in the survey. The study took place at the end of the academic year before final examinations of students and lasted for 3 months. Information about students' characteristics is provided in Table 2.

The present study applied all quality dimensions of business quality tool ServQual model, adapted to higher education to measure the quality of provided administrative services. The innovativeness of this study in applying the ServQual model to identify service quality from perspectives of students is that mean value of 22 items in all dimensions of expectation part have already been scored as '5', '4.5' '4' and '3.5', defined as ideal values. Students assessed the provided service quality by assessing already perceived services (please refer to Table 3).

Table 2 Demographic backgrounds of respondents ($n = 322$)

Features	Percent	Number
Gender		
Male	31,1	100
Female	68,9	222
Age		
17–18	8,4	27
19–20	63,7	205
21–22	23,9	77
23 than older	4	13
Nationality		
Kazakh	90.4	291
Russian	4.7	15
Other	4.9	16
Major		
Humanitarian sciences	39,8	128
Natural sciences	52,2	168
Art	3,4	11
Medicine	0,3	1
Technical sciences and technologies	3,7	12
Economics, business and law	0,6	2

Source Adapted by Author [27]

4 Data Analyses and Results

The demographic background of the survey in Table 2 shows that only one-third of respondents were males. As for the age distribution, the majority of participants belong to the age group of 19–20 (63.7%), and just under one quarter belong to the age group of 21–22. As for the nationality of respondents, Kazakh representatives numbered to just over 90% of the total, and only less than one-twentieth compose Russian and other nationalities. In the same, manner, the research survey had the following information about major of respondents. More than half of respondents majored in natural studies, whereas more than two-fifth specialize in humanitarian sciences. Regarding the rest field of studies, representatives of art and technical sciences were less than one twenty fifths of the total in comparison with medicine and business, law, economics, which composed less than 1% of the total.

The study attempted to identify the quality gap in services provided by administrations of university and schools. The survey was based on the identification of the level of satisfaction with the administration of the university. As a measurement tool, Cronbach's α was applied for identifying reliability consistency of the survey. The reason for using this method was to identify if the designed questionnaire was

Table 3 The dimensions of quality—ServQual model

Dimensions	Variables
Tangibility	Q1—Availability of materials and necessary modern technologies designed for learning
	Q2—Modernity and attractiveness of the university and the faculty
	Q3—University staff and faculty are well dressed and neat
	Q4—The university has special classrooms for practical, laboratory classes and individual work of students
	Q5—Availability of facilities to provide social services and to fully meet the diverse needs of students (canteens, accommodation, sports infrastructure, educational, medical, legal, financial and cultural services)
Reliability	Q6—An opportunity to contact with faculty staff and managers at any time to get a satisfactory answer on questions
	Q7—The sincere interest of faculty staff and university administration to solve a problem
	Q8—Faculty staff immediately meeting students' requirements and needs
	Q9—Requests and requirements of students are processed and resolved in a timely and efficient manner
	Q10—Information of certain events in time (exams, consultations, conferences, seminars, etc.), and fulfilment of students' requests
Responsiveness	Q11—A prompt respond of faculty to students' inquiries
	Q12—Feeling of support and assistance
	Q13—Willingness and openness of faculty staff to answer questions and requirements
	Q14—Convenience of Student Office hours to meet students' needs
Assurance	Q15—University and faculty staff are always polite to students
	Q16—Students trust faculty staff
	Q17—Feeling of safety in the learning process
	Q18—Lectures are given by teachers who are experts in their fields and possess the necessary knowledge and skills
Empathy	Q19—Individual attention of faculty and university staff to each student
	Q20—Equal treatment and respect
	Q21—Taking student interests into account
	Q22—Understanding the specific needs of students

Source Adapted and developed by Author [14]

accurately measuring the variable of interest and to identify the level of satisfaction via reliability indicators. The standardized coefficient of Cronbach alpha α_{st} is calculated in the following way:

$$\frac{N \cdot \overline{r}}{1 + (N - 1) \cdot \overline{r}} \tag{1}$$

Table 4 Reliability statistics

Total items/questions/components	22
the sum of the item variances	25,12
the variance of total scores	246,89
Cronbach's a	0,941

Table 5 Reliability scores of the research constructs

Variables	No of Items	Cronbach's Alpha scores
Tangibles	5	0,752
Reliability	5	0,847
Responsiveness	4	0,823
Assurance	4	0,799
Empathy	4	0,872

Source Authors' own reproduction

where N is the number of studied items, and $\overline{r}$ determines the average correlation coefficient between the items. Besides, the coefficient can be calculated by the following formula:

$$\alpha_{st} = \frac{N}{N-1}\left(1 - \frac{\sum_{i=1}^{N} \sigma_{Y_i}^2}{\sigma_X^2}\right) \quad (2)$$

where N measures the number of components under study, σ_X^2 is the standard deviation of all the sets studied, and $\sigma_{Y_i}^2$ is the standard deviation of an individual item.

In this research, the second formula was applied.

The results in Table 4 illustrates that the total reliability of the measurement scale had an alpha coefficient of 0.941, which demonstrates excellent consistency in compliance with Cronbach's alpha internal consistency Table 4 [28].

In addition, the present paper calculated reliability consistency for all five dimensions of SERVQUAL survey each. The findings given in Table 5 demonstrate acceptable consistency for all dimensions.

4.1 *Statistical Data Processing Methods*

Wilcoxon's sign rank test applied in the study is a nonparametric, one-sample criterion, does not require a normal distribution of the compared populations. It compares the median with the hypothetical median and determines whether the gap between the actual and hypothetical median is random or non-random. The null hypothesis of this test is that the values of perceptions are not statistically significantly different from the hypothetical median or estimation. An alternative hypothesis, respectively, is the presence of a statistically significant difference between the assessment of perceptions and hypothetical expectations. The null hypothesis can be rejected, with an error level of less than 5%, if the probability of obtaining such a gap is less than 5%, then the initial hypothesis is rejected, and this means that such a gap is not random.

From the beginning, the aim of this paper was an attempt to testify the applicability of the ServQual model in the context of Kazakh national university and to identify if the tool is a useful approach to determine the missing gap between expected and perceived values of provided services based on already proposed by Parasuraman y[29] five dimensions. The study attempted to identify the gap through already measuring mean expectation scores ranging from four different acceptable scores '5', '4.5', '4' to '3.5'. The essence of the applied method was to compare absolute values of ranks' shifts in positive or negative directions. To do this, first, all absolute values of the shifts have been ranked, and then the ranks have been summed. If shifts are random, then the sums of their ranks will be approximately equal. If the frequency of shift in one direction is greater, the sum of absolute value ranks will be lower. If the sum of shifts is not equal, then the difference between expected and perceived values is significant (Table 6).

Wilcoxon's sign test was conducted using the IBM SPSS Statistics (Statistical Package for the Social Sciences) programme, version 23 is a computer programme for statistical data processing, one of the market leaders in the field of commercial statistical products designed for applied research in the social sciences.

The mean score of the service quality perceptions in case of the ideal score '5' showed all 100% significant results of cases, i.e. differences that the gap between the actual and expected perceptions can be considered proven and not random, but regular. The mean hypothetical score '4.5' demonstrated the same results as the mean score for "5"—the difference between the actual and expected perceptions is statistically significant at 100% Tables 7 and 8.

Only in case of the mean value is '4', three variables showed an insignificant result related to tangibles (Q3—the highest mean score 4,2), reliability (Q6—4,0) and assurance (Q18—4,1) dimensions in perception aspect (university and faculty staff are well dressed and neat / an opportunity at any time, to contact faculty staff and managers and to get a satisfactory answer on questions / Students trust faculty staff) (Table 9).

With a median value of 3.5 points, 21 out of 22 variables showed an insignificant result (95%). Thus, observations demonstrate a relationship between the value of the hypothetical median and the share of significant differences. The lower the expectations are the less difference with the actual perception of provided services (Table 10). Results showed that the students have relatively high perception scores related to *all four* dimensions, except for *tangibility*, where Q1 (*Availability of materials and necessary modern technologies designed for learning)* had the most negative quality gap.

To summarize the obtained results, the study assessed all 22 variables of ServQual dimensions as ideal and perfect variables, which encompass all aspects of services

Table 6 Summary of Test results by the criterion of Wilcoxon's iconic ranks when comparing perceptions with hypothetical expectation in four different mean scores

Hypothesis test summary

Mean score	Items	Significance	Decision
5.0	Q1–Q22	0,000	Reject the null hypothesis
4.5	Q1–Q22	0,000	Reject the null hypothesis
4.0	Q1–Q5, Q7–9, Q11–16, Q19–22	0,000	Reject the null hypothesis
	Q6	0,914	Retain the null hypothesis
	Q10	0,020	Reject the null hypothesis
	Q17	0,016	Reject the null hypothesis
	Q18	0,070	Retain the null hypothesis
3.5	Q1, Q3, Q5, Q6, Q10–13, Q17–18	0.000	Reject the null hypothesis
	Q2	0,740	Retain the null hypothesis
	Q4	0,107	Retain the null hypothesis
	Q7	0,090	Retain the null hypothesis
	Q8	0,768	Retain the null hypothesis
	Q9	0,224	Retain the null hypothesis
	Q14	0,417	Retain the null hypothesis
	Q15	0,129	Retain the null hypothesis
	Q16	0,009	Reject the null hypothesis
	Q19	0,412	Retain the null hypothesis
	Q20	0,002	Reject the null hypothesis
	Q21	0,094	Retain the null hypothesis
	Q22	0,951	Retain the null hypothesis

Asymptotic significances are displayed. The significance level is 0,5
Source Authors' own reproduction

provided by the university administration. Accordingly, the results of the research can serve as a guideline for university authorities to develop a quality management tool using the proposed four different assessment of the gap quality.

The significant values in perceptions were obtained when the study had '4' and 3,5' mean score for expectations, while there was no significant value in case of '5' and '4.5' mean score of expected dimensions. Also, there were significant differences between perceptions and expectations of students in all five dimensions (Table 11).

Following, we can claim that the obtained results of the study can help academics and administrators allocate their financial, material and human resources accordingly and figure out the most important dimensions of service quality, subject to improvement.

Table 7 Mean scores of expectations and perceptions, and service quality gap in each dimension (mean score = 5)

Quality dimension	Question	Expectations	Perception	Quality Gap	One-sample Wilcoxon signed rank test	
					Sig.	Decision
Tangibles	Q1	5	3,2	−1,8	0,000	Reject the null hypothesis
	Q2	5	3,5	−1,5	0,000	Reject the null hypothesis
	Q3	5	4,2	−0,8	0,000	Reject the null hypothesis
	Q4	5	3,6	−1,4	0,000	Reject the null hypothesis
	Q5	5	3,7	−1,3	0,000	Reject the null hypothesis
Reliability	Q6	5	4,0	−1,0	0,000	Reject the null hypothesis
	Q7	5	3,5	−1,5	0,000	Reject the null hypothesis
	Q8	5	3,5	−1,5	0,000	Reject the null hypothesis
	Q9	5	3,4	−1,6	0,000	Reject the null hypothesis
	Q10	5	3,8	−1,2	0,000	Reject the null hypothesis
Responsiveness	Q11	5	3,7	−1,3	0,000	Reject the null hypothesis
	Q12	5	3,7	−1,3	0,000	Reject the null hypothesis
	Q13	5	3,7	−1,3	0,000	Reject the null hypothesis

(continued)

Table 7 (continued)

Quality dimension	Question	Expectations	Perception	Quality Gap	One-sample Wilcoxon signed rank test	
					Sig.	Decision
	Q14	5	3,5	−1,5	0,000	Reject the null hypothesis
Assurance	Q15	5	3,6	−1,4	0,000	Reject the null hypothesis
	Q16	5	3,6	−1,4	0,000	Reject the null hypothesis
	Q17	5	3,8	−1,2	0,000	Reject the null hypothesis
	Q18	5	4,1	−0,9	0,000	Reject the null hypothesis
Empathy	Q19	5	3,5	−1,5	0,000	Reject the null hypothesis
	Q20	5	3,6	−1,4	0,000	Reject the null hypothesis
	Q21	5	3,5	−1,5	0,000	Reject the null hypothesis
	Q22	5	3,4	−1,6	0,000	Reject the null hypothesis

Source Authors' own reproduction

5 Conclusion

In this research paper, a perceived service quality measurement scale has been applied to determine the level of student satisfaction with provided services by university/faculty management. The findings of the research show that the SERVQual scale is as a measurement tool successfully applicable in higher education to determine the quality of education, which in turn ensures reliability and accuracy.

First of all, in terms of managerial application, the current paper has demonstrated the applicability of the well-known service quality measurement tool in case of Kazakhstani university, which can be used by practitioners, university managers in developing a quality management practices. Secondly, the findings can be found as significant guidelines for university management in designing quality management

Table 8 Mean scores of expectations and perceptions, and service quality gap in each dimension (mean score = 4,5)

Quality dimension	Question	Expectations	Perception	Quality Gap	One-sample Wilcoxon signed rank test	
					Sig.	Decision
Tangibles	Q1	4,5	3,2	−1,3	0,000	Reject the null hypothesis
	Q2	4,5	3,5	−1,0	0,000	Reject the null hypothesis
	Q3	4,5	4,2	−0,3	0,000	Reject the null hypothesis
	Q4	4,5	3,6	−0,9	0,000	Reject the null hypothesis
	Q5	4,5	3,7	−0,8	0,000	Reject the null hypothesis
Reliability	Q6	4,5	4,0	−0,5	0,000	Reject the null hypothesis
	Q7	4,5	3,5	−1,0	0,000	Reject the null hypothesis
	Q8	4,5	3,5	−1,0	0,000	Reject the null hypothesis
	Q9	4,5	3,4	−1,1	0,000	Reject the null hypothesis
	Q10	4,5	3,8	−0,7	0,000	Reject the null hypothesis
Responsiveness	Q11	4,5	3,7	−0,8	0,000	Reject the null hypothesis
	Q12	4,5	3,7	−0,8	0,000	Reject the null hypothesis
	Q13	4,5	3,7	−0,8	0,000	Reject the null hypothesis

(continued)

Table 8 (continued)

Quality dimension	Question	Expectations	Perception	Quality Gap	One-sample Wilcoxon signed rank test	
					Sig.	Decision
	Q14	4,5	3,5	−1,0	0,000	Reject the null hypothesis
Assurance	Q15	4,5	3,6	−0,9	0,000	Reject the null hypothesis
	Q16	4,5	3,6	−0,9	0,000	Reject the null hypothesis
	Q17	4,5	3,8	−0,7	0,000	Reject the null hypothesis
	Q18	4,5	4,1	−0,4	0,000	Reject the null hypothesis
Empathy	Q19	4,5	3,5	−1,0	0,000	Reject the null hypothesis
	Q20	4,5	3,6	−0,9	0,000	Reject the null hypothesis
	Q21	4,5	3,5	−1,0	0,000	Reject the null hypothesis
	Q22	4,5	3,4	−1,1	0,000	Reject the null hypothesis

Source Authors' own reproduction

procedures in accordance with students' needs. Thirdly, the results of the research illustrate that all five dimensions of the SERVQual encompass all required tangible and intangible features of university service. Apart from, adequate physical facilities of the university, dimensions of the measurement scale cover relationship between students and administration, attitude of university/faculty management in providing service to students, as well as the feeling of confidence, trust and loyalty between all key actors integrated. The study suggests to university managers to develop quality management system, which encompasses all aspects of service provided to students taking into account the observed educational quality gap.

The main purpose of the paper was to provide a practical framework on applicability of the service quality tool in Kazakhstani HE. The originality of this study is that a definite mean score of expectations in all five dimensions has been proposed

Table 9 Mean scores of expectations and perceptions, and service quality gap in each dimension (mean score = 4)

Quality dimension	Question	Expectations	Perception	Quality gap	One-sample Wilcoxon signed rank test	
					Sig.	Decision
Tangibles	Q1	4	3,2	−0,8	0,000	Reject the null hypothesis
	Q2	4	3,5	−0,5	0,000	Reject the null hypothesis
	Q3	4	4,2	0,2	0,000	Retain the null hypothesis
	Q4	4	3,6	−0,4	0,000	Reject the null hypothesis
	Q5	4	3,7	−0,3	0,000	Reject the null hypothesis
Reliability	Q6	4	4,0	0,0	0,914	Retain the null hypothesis
	Q7	4	3,5	−0,5	0,000	Reject the null hypothesis
	Q8	4	3,5	−0,5	0,000	Reject the null hypothesis
	Q9	4	3,4	−0,6	0,000	Reject the null hypothesis
	Q10	4	3,8	−0,2	0,020	Reject the null hypothesis
Responsiveness	Q11	4	3,7	−0,3	0,000	Reject the null hypothesis
	Q12	4	3,7	−0,3	0,000	Reject the null hypothesis
	Q13	4	3,7	−0,3	0,000	Reject the null hypothesis

(continued)

Table 9 (continued)

Quality dimension	Question	Expectations	Perception	Quality gap	One-sample Wilcoxon signed rank test	
					Sig.	Decision
	Q14	4	3,5	−0,5	0,000	Reject the null hypothesis
Assurance	Q15	4	3,6	−0,4	0,000	Reject the null hypothesis
	Q16	4	3,6	−0,4	0,000	Reject the null hypothesis
	Q17	4	3,8	−0,2	0,016	Reject the null hypothesis
	Q18	4	4,1	0,1	0,070	Retain the null hypothesis
Empathy	Q19	4	3,5	−0,5	0,000	Reject the null hypothesis
	Q20	4	3,6	−0,4	0,000	Reject the null hypothesis
	Q21	4	3,5	−0,5	0,000	Reject the null hypothesis
	Q22	4	3,4	−0,6	0,000	Reject the null hypothesis

Source Authors' own reproduction

ranging from 5 to 3.5 scores. The findings of the current study could be a useful guideline for university managers in terms of managerial implications to enhance service quality in the universities. The outcomes of the research demonstrate that the higher the expectation score is the less satisfaction level of students with the provided administrative services.

The observations showed a gap between the students' perceptions and expectations for all five dimensions. Indeed, all scores for perceived dimensions were lower than the expectation scores in case of the high mean value of expectations, which indicates a need for quality improvements of provided service by university management and faculty staff.

Table 10 Mean scores of expectations and perceptions, and service quality gap in each dimension of the quality of educational services (mean score = 3.5)

Quality dimension	Question	Expectations	Perception	Quality gap	One-sample Wilcoxon signed rank test	
					Sig.	Decision
Tangibles	Q1	3,5	3,2	−0,339	0,000	Reject the null hypothesis
	Q2	3,5	3,5	−0,047	0,740	Retain the null hypothesis
	Q3	3,5	4,2	0,652	0,000	Reject the null hypothesis
	Q4	3,5	3,6	0,056	0,107	Retain the null hypothesis
	Q5	3,5	3,7	0,193	0,000	Reject the null hypothesis
Reliability	Q6	3,5	4,0	0,500	0,000	Reject the null hypothesis
	Q7	3,5	3,5	0,050	0,090	Retain the null hypothesis
	Q8	3,5	3,5	−0,034	0,768	Retain the null hypothesis
	Q9	3,5	3,4	−0,130	0,224	Retain the null hypothesis
	Q10	3,5	3,8	0,345	0,000	Reject the null hypothesis
Responsiveness	Q11	3,5	3,7	0,189	0,000	Reject the null hypothesis
	Q12	3,5	3,7	0,193	0,000	Reject the null hypothesis
	Q13	3,5	3,7	0,214	0,000	Reject the null hypothesis

(continued)

Table 10 (continued)

Quality dimension	Question	Expectations	Perception	Quality gap	One-sample Wilcoxon signed rank test	
					Sig.	Decision
	Q14	3,5	3,5	−0,012	0,417	Retain the null hypothesis
Assurance	Q15	3,5	3,6	0,062	0,129	Retain the null hypothesis
	Q16	3,5	3,6	0,090	0,009	Reject the null hypothesis
	Q17	3,5	3,8	0,342	0,000	Reject the null hypothesis
	Q18	3,5	4,1	0,571	0,000	Reject the null hypothesis
Empathy	Q19	3,5	3,5	0,009	0,412	Retain the null hypothesis
	Q20	3,5	3,6	0,130	0,002	Reject the null hypothesis
	Q21	3,5	3,5	0,037	0,094	Retain the null hypothesis
	Q22	3,5	3,4	−0,075	0,951	Retain the null hypothesis

Source Authors' own reproduction

Table 11 The comparison of "expectation" scores in absolute values

Mean score	5	4.5	4	3.5
Negative quality gap	22	22	20	1
Positive quality gap	0	0	3 (13%)	22 (5%)

Source Authors' own reproduction

6 Limitation and Future Research

This study has a certain number of limitations. In addition, it leads to further study of business quality measurement tools applicable in higher education.

First of all, the sample in this research is limited only to population of al-Farabi Kazakh National University in Kazakhstan. The current research leads to further study and application of the ServQual in other Kazakhstani HEIs with different profile and mission.

Secondly, since this study only concentrated on undergraduates, future research in determining service quality should be carried out among postgraduate students, mainly among Master's and PhD degree students. Peculiarities, education and demographic background, attitude and worldwide view of postgraduate students significantly differ from undergraduates, thus findings of a new study will be exponential and worthy of attention for academics and managers of higher education.

Overall, the study presents an exceptional assessment of the gap between expected and perceived values of quality service dimensions, by prior defining mean scores of expectations as ideal.

References

1. Zhao X.: A Review on service quality and student satisfaction of higher education. In: Luo J. (eds.) Soft Computing in Information Communication Technology. Advances in Intelligent and Soft Computing, vol. **158**. Springer, Berlin, Heidelberg (2012)
2. Lewis, B.R., Smith, A.M.: Customer Care in the Service Sector: The Customer's Perspective, Financial Services Research Centre, Manchester School of Management, Manchester (1989)
3. Lewis, R., Booms, B.: The marketing aspect of service quality. Emerging perspectives on services marketing. American Marketing Association, Illinois, USA (1983)
4. Abdullah, F.: The development of HEdPERF: a new measuring instrument of service quality for the higher education sector. Int. J. Consum. Stud. **30**(6), 569–581 (2006)
5. Abdullah, F.: Measuring service quality in higher education: three instruments compared. Int. J. Res. Method Educ. **29**(1), 71–89 (2006)
6. Parsu Parasuraman, A., Valarie, A., Zeithaml, L.L.: Berry: ServQual: a Mutiple-item scale for measuring consumer perceptions of service quality. J. Retail. **64**(1), 12–40 (1988)
7. Harvey, L.: Editorial: student feedback. Qual. High. Educ. **9**(1), 3–20 (2003)
8. Quinn, A., Gina, L., Larsen, P.: Service quality in higher education. Total Qual. Manag. Bus. Excell. **20**(2), 139–152 (2009)
9. Wang, R.: Study on the cause-effect relationship of the service quality evaluation criteria of portal sites. African J. Bus. Manage. **4**(10), 2432–2444 (2011)
10. Abouchedid, K., Ramzi, N.: Nasser: assuring quality service in higher education: registration and advising attitudes in a private university in Lebanon. Qual. Assur. Educ. **10**(4), 198–206 (2002)
11. Marrs, M., Gajos, R., Pinar, M.: Utilizing Brand Audit to Develop a University Brand: A Case Study. ASBBS Annual Conference, pp. 964–975. Proceedings of ASBBS, Las Vegas (2011)
12. Soni S., Govender K.: The Relationship between service quality dimensions and brand equity: higher education students' perceptions. J. Manag. Bus. Administration. Cent.L Eur. **26**(3), 71–87 (2018)
13. Roland, K.: Yeo & Jessica Li: Beyond SERVQUAL: the competitive forces of higher education in Singapore. Total Qual. Manag. Bus. Excell. **25**(1–2), 95–123 (2013). https://doi.org/10.1080/14783363.2011.637802
14. Soares, M.C., Novaski, O., Anholon, R.: Servqual model applied to higher education public administrative services. Braz. J. Oper. Prod. Manag. **14**(3), 338–349 (2017)
15. Cristina, C.P., Le´vy-Mangin, J.P., Novo-Corti, I.: Perceived quality in higher edu¬cation: an empirical study. Mark. Intell. Plan. **31**(6), 601–619 (2013)

16. Robinson, S.: Measuring service quality: current thinking and future requirements. Mark. Intell. Plan. **17**(1), 21–32 (1999)
17. Waugh, R.F.: Academic perceptions of administrative quality at universities. J. Educ. Adm. **40**(2), 172–188 (2002)
18. Parasuraman, P.A., Zeithaml, V.A., Berry, L.L.: A conceptual model of service quality and its implications for future research. J. Market. **49**(Fall) 41–50 (1985)
19. Brysland, A., Curry, A.: Service improvements in public services using SERVQUAL. Manag. Serv. Qual. **11**(6), 389–401 (2001)
20. Al-Alak, B., Alnaser, A.: Assessing the relationship between higher education service quality dimensions and student satisfaction. Aust. J. Basic Appl. Sci. **6**(1), 56–164 (2012)
21. Angell, R.J., Hefferman, T.W., Megicks, P.: Service quality in postgraduate education. Qual. Assur. Educ. **26**(3), 236–254 (2008)
22. Hussain, K., Birol, C.: The assessment of non-academic and academic service quality in higher education. Eurasian J. Educ. Res. **1**(42), 95–116 (2011)
23. Yousapronpaiboon, K.: SERVQUAL: measuring higher education service quality in Thailand. Procedia. Soc. Behav. Sci. **116**(2014), 1088–1095 (2013)
24. Khomonenko, A.D., Dashonok, V.L., Ivanova, K.A., Kassymova, D. T.: Approach to processing of data from social networks for detecting public opinion on quality of educational services. In: 20th IEEE International Conference on Soft Computing and Measurements (SCM). St Petersburg, Russia, IEEE (2017)
25. Kerimkulova S., Kuzhabekova A.: Quality assurance in higher education of Kazakhstan: a review of the system and issues. In: rise Of quality assurance in asian higher education, pp. 87–108 (2017)
26. Nadiri, H., Kandampullyb, J., Hussain, K.: Students' perceptions of service quality in higher education. Total Qual. Manag. Bus. Excell. **20**(5), 523–535 (2009)
27. Sfakianaki, E., Rizos, S., Kakouris, A.: Quality of administrative services in higher education institutions and customer satisfaction: a servQual application. In: total quality management (TQM) concepts, implementation and applications. Nova science publishers, New York (2019)
28. Tavakol, M., Dennick, R.: Making sense of Cronbach's alpha. Int. J. Med. Educ. **2**, 53–55 (2011)
29. Alves, A., Vieira, A.: SERVQUAL as a marketing instrument to measure service quality in higher education institutions. In: the second international conference: product management challenges of the future (2006)
30. Akareem, H.S., Hossain, S.S.: Determinants of education quality: what makes students' perception different? Open Rev. Educ. Res. 52–67 (2016)
31. Bradley, R.: Barnes: analysing service quality: the case of post-graduate chinese students. Total Qual. Manag. **18**(3), 313–331 (2007)
32. Railya, B.: Galeeva: SERVQUAL application and adaptation for educational service quality assessments in Russian higher education. Qual. Assur. Educ. **24**(3), 329–348 (2016)
33. Vajda, B.K., Farkas, G., Málovics, E.: Student evaluations of training and lecture courses: development of the COURSEQUAL method. Int. Rev. Public Nonprofit Mark. **12**(1), 79–88 (2015)
34. Voss, R., Gruber, Th., Szmigin, I.: Service quality in higher education: the role of student expectations. J. Bus. Res. **60**(9), 949–959 (2007)
35. Thuy, V.T.N., Thao, H.D.P.: Impact of students' experiences on brand image perception: the case of Vietnamese higher education. Int. Rev. Public Nonprofit Mark **14**(2), 217–251 (2017)
36. Legal information system of Regulatory Legal Acts of the Republic of Kazakhstan.: Law of the Republic of Kazakhstan dated November 13, 2015 No. 398-V ЗRK "On amendments and additions to some legislative acts of the Republic of Kazakhstan on education". http://adilet.zan.kz/rus/docs/Z1500000398#z6 (2015). Last Accessed 06 Sept 2021

Education Industry—New Effective Reality

Tatiana Bondarenko, Aleksey Bolvachev, and Natalia Bondarenko

Abstract Now artificial intelligence is becoming a great part of our normal life. Humanity is surrounded by this technology by automatic parking systems and smart sensors to obtain photographs and personal assistance. Similarly, there is the development of modern technology and artificial intelligence in education, where traditional methods are radically changing. The academic world is becoming more convenient and personalized thanks to the numerous applications of modern technologies and artificial intelligence for education. Smart technologies in education include the facility to create and use virtual content (video conferencing, video lectures), systems of artificial intelligence used a traditional curriculum to create customized textbooks for specific subjects, the textbooks digitized and created new training interfaces to facilitate the educational process for all ages of students. It changed the way of teaching people, as teaching materials become available to all through modern devices and computers. In addition, this approach allows providing the automation process of some administrative tasks. Modern technology can be used directly in the process of education, as well as for automation of the evaluation task, involving several tests, and software vendors offer the best methods of assessment written answers and essays.

Keywords Smart technologies in education · Modern technologies · Artificial intelligence · Facilitation of the educational process · Platforms online learning

---

T. Bondarenko (✉) · A. Bolvachev
Plekhanov Russian University of Economics, 36, Stremyanny lane, 117997 Moscow, Russian Federation
e-mail: t.g.bondarenko@gmail.com

N. Bondarenko
Tula State Lev Tolstoy Pedagogical University, 125, Lenin Avenue, 300026 Tula, Russian Federation

G. Mutanov and A. Serikbekuly (eds.), *Digital Transformation in Sustainable Value Chains and Innovative Infrastructures*, Studies in Systems, Decision and Control 443,
https://doi.org/10.1007/978-3-031-07067-9_5

1 Introduction

Digital technology and artificial intelligence (AI) have changed the way of teaching and learning. Knowledge is no longer limited to the physical boundaries of a traditional university or college audience. In modern conditions of learning, access to unlimited information is on the fingertips of students and their devices. Teachers can rely on these technologies and use them to become a co-designer of learning, and to develop the learning process together with their students, supporting and guiding students with the help of digital technologies. It means that they provide an opportunity to study independently. In this sense, every teacher can become a student and vice versa.

The modern world and the integration processes in all areas of life show that the directions of development of both scientific and cultural space are pointed by the conditions of both globalization and glocalization. It requires to par constant attention to the problems of adaptability of education (in a broad sense) to changing circumstances. According to some scientists (Zabrodina et al. [1], Shutenko and Shutenko [2] have emphasized in their works) the intensification of global integration processes and Russia's growing immersion into this process reveal the challenge of training of a new generation of analysts and researchers and managers with a mentality of professionals. And these professionals have to think globally and always be able to identify and effectively use the opportunities of a rapidly changing business environment and be prepared to adequately and timely respond to its challenges.

Effective education is education that provides a solid basis (the basis of knowledge and search methods for their acquisition and expansion) and can be used in the future to solve the tasks through the development of meta-subject communications in new, completely unrelated situations. Integration processes make it possible to reconsider the concept of "learning" not as a theory, the boundaries of which are defined by the academic framework, where students' achievements are and their ability to recall previously obtained information, transmitted details, the sequence of performing any traditional steps to solve conditional problems is evaluated. In modern conditions, the transition to competencies from the skills and ability to remember, act and follow the standards set by the standards requires the development of intersubject and metasubject communications, the ability to jump between related sciences and drag algorithmic knowledge and skills in solving problems in intersubject areas. It is such a strategy for the development of teaching and learning that should increase the competence of students that is highlighting the specific attributes of a modern system of teaching and learning methods and techniques, which are accompanied by a revolutionary round of development of artificial intelligence and the digital economy [3].

The traditional model of education, is able only to train people seeking job-seekers, but it does not take into account any of these paradoxical crises [4]. In this fast-paced world of change, knowledge is now the main driver, and the creation and development of online courses and platforms are more likely to address the lack of talent and mass youth unemployment. The new economy needs specialists who have adaptive experience in order to be innovative, flexible and creative in various contexts [5].

The mandatory attributes of modern education in the context of the integration process are:

- understanding the process of how students learn (ensuring timely, specific feedback from the student, tracking and recording progress, starting with the student's self-assessment);
- compliance with the principles of critical thinking of teaching and learning (support for the acquisition of knowledge, skills of professional behavior in the learning process, ensuring timely, specific feedback from students);
- facilitation, not control process of learning (individualization of the learning path, the expected and timely increase in complexity as the student progresses throughout the program along the "proposed" path of the educational process) [6];
- effective assistance to the competency-based approach through building relevant regulatory and instructional documents of competency-based educational programs, including those implemented in various formats, for example, through open educational resources, using video materials, recorded lectures, and through project training. Many foreign programs today are intended for students who have chosen an individual educational path, realizing them at their own pace, rather than at the pace dictated by semesters or credit hours [7]. This approach means that motivated and effective students can complete a dedicated educational level in less time, and vice versa.

The process of studying the disciplines of the chosen educational program is aimed at the formation of many competencies. It allows us to talk about such final holistic results as the ability to use the information (knowledge, skills) to make decisions. This aim is achieved through the use of the following forms and training methods that contribute to the formation and development of competencies as a lecture, essay, essay, presentation, seminar, case study, test task, didactic game, test, analytical reports, informational reviews, practice, research work, as well as various types of independent work of students on the instructions of the teacher. It means that it is necessary to use both active and interactive forms of work simultaneously.

The integration processes taking place in society and affecting the scientific and cultural space—reducing the autonomy of universities, a new world view, moving away from the industrial model of society and moving to a new model of production and logistics systems, the emergence of a map of promising professions—it all forms identical requirements for educational programs, especially individual educational program, and becomes the new institutional core of training and education.

2 Results

Artificial intelligence can change the learning process and affect education. AI-based programs are created specifically to complement the learning process. Of course, AI will not replace a completely "lively" classroom process, but it will help to use relevant practices, so that, for example, by tracking students' answers, they can be warned that they need additional consultation or explanation on issued but poorly digestible materials. A similar model of teaching and learning will provide real-time feedback on student progress; will assist teachers in creating textbooks and exercises adapted to the needs of their specific courses and students; will allow you to work in a simulated and game experience of a virtual type (using the problematic approach), which will quickly involve students in the educational process [10].

Thus, AI (for example, artificial scientific advisers) has the potential to improve online learning using adaptive software for education and modeling of an intuitive instant response to the process of interaction with students. As technology becomes more advanced, these AI advisors will better understand and respond to the intricacies of linguistics, gestures, and tones that vary among all students. AI has tremendous potential for developing creative needs and non-formal learning (see Table 1).

Today we can say that VUCA as a new environment for managing knowledge, talents and personnel is the most appropriate tool for integrating the educational process into integration processes in society (see Fig. 1).

Massive open online courses (MOOCs)—provide an opportunity, in combination with an individual personal choice, to take meaningful actions to form one's own learning path (discipline, duration and intensity of the course, manner of presenting the material, level of material submitted) and see the results of your decisions. The emergence and development of platforms for online education can be considered as a catalyst for changes or transformations of education throughout the life of each individual person. In this context, the platform for students is aimed at transferring responsibility for education from the teacher to the student, allowing him to gain understanding, mental faculty and ability to participate in the learning process and take measures to intervene in the educational process in order to be able to influence his or her results in necessary point in time and in accordance with personal professional or personal requests [8, 9].

Online learning platforms are the driving forces that enable talented people to grow. Active children turn into active adults who invest in knowledge today to use it in a crisis situation. In addition, such online courses have a social effect, providing (facilitating) access to obtaining the necessary information. Reducing the demand for a particular competency (or even a profession), unemployment—it leads to personal poverty, psychological trauma and social unrest. In this case, social inequality flourishes, as a lack of talent leads to an increase in the income of highly skilled workers, which, in turn, leads to an even greater income gap.

The effect of the presence of artificial intelligence in education can be graphically represented (see Fig. 2).

Table 1 Generalized results of a comparative analysis of the traditional education system and the innovation system using Artificial intelligence

Comparison criteria	Traditional learning approach	AI learning approach
Goal	activity information transfer	organization of cognitive activity
Learning Content	information on professional knowledge and experience gained, integrated professional knowledge	creative work experience, professional development of personality
Principles	science, accessibility, collective cognitive activity	unity of cognitive and research activities
Methods	reproductive, heuristic, research	problem statement
Mechanisms of activity	actively receiving and remembering information about activities and experience	creative goal-setting interaction; modeling; scientific substantiation, hypothesis and hypothesis testing
Types of cognitive activity	obtaining information about the experience gained by mankind	study of the experience gained by mankind in a professional environment
Teacher's role	activity information transmitter, organizer of cognitive activities of masters	translator of the experience of scientific and professional relations
Management level	high: information management	medium: information management
Learning tools	educational and scientific literature of the classical type, information media as technical teaching aids	information media of any type (primarily mobile and remote)

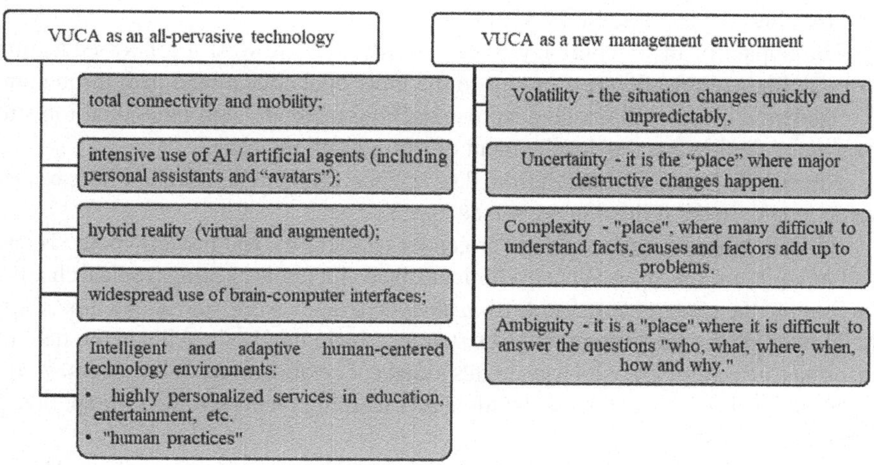

Fig. 1 Technology and Education: VUCA as the norm

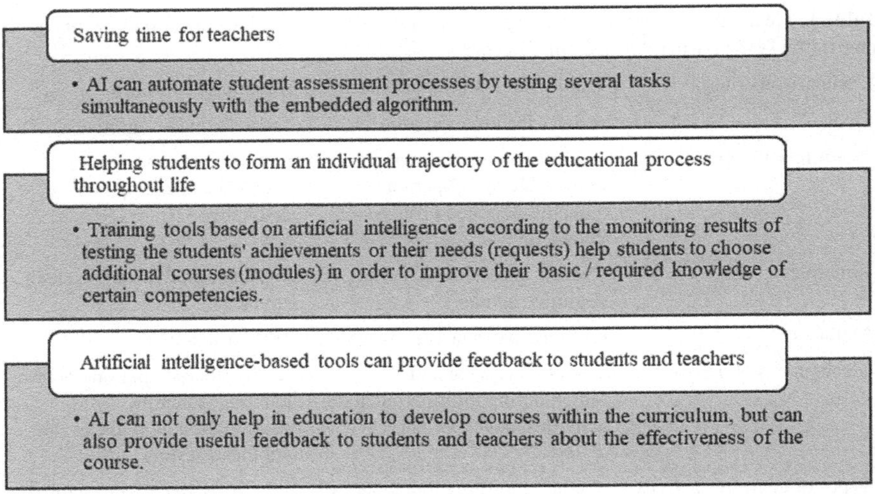

Fig. 2 The effect of the artificial intelligence's presence in education

Thus, almost all industries use AI for their development, and the education sector, as one of the most important sectors, must also adapt to these changes.

AI will become more and more "intellectual"—this means that the data on the work of students will be processed statistically and in the future will help to form a personal educational trajectory for each student depending on the personal results and the progress of training and requests.

A balanced combination of human and artificial intelligence [11] will help translate the best teaching practices to meet the individual needs of each student.

A study published by eSchool News shows that by 2021, the use of AI in education and training will increase by 47.5% [12].

The impact of this technology will be felt from the lowest levels of education through higher education institutions to the individual educational path throughout the student's life. It will create adaptive teaching methods with individual tools to facilitate learning and the development of self-education needs.

AI can help humanity benefit from data science, which will make our lives easier, for example, in the following areas (see Fig. 3).

Education has no territorial, age or gender restrictions, and modern methods and technologies in education can help eliminate them. Technology brings radical change, facilitating the study of any course from anywhere in the world and at any time. Artificial intelligence-based training gives students, including basic skills in the field of information technology, which generates a chain of creating more online courses with different levels of content, while building self-learning processes and the experience transfer processes.

It is especially necessary to note the role of smartphones and tablets, which at times strengthened and accelerated the integration processes of modern society [13, 21].

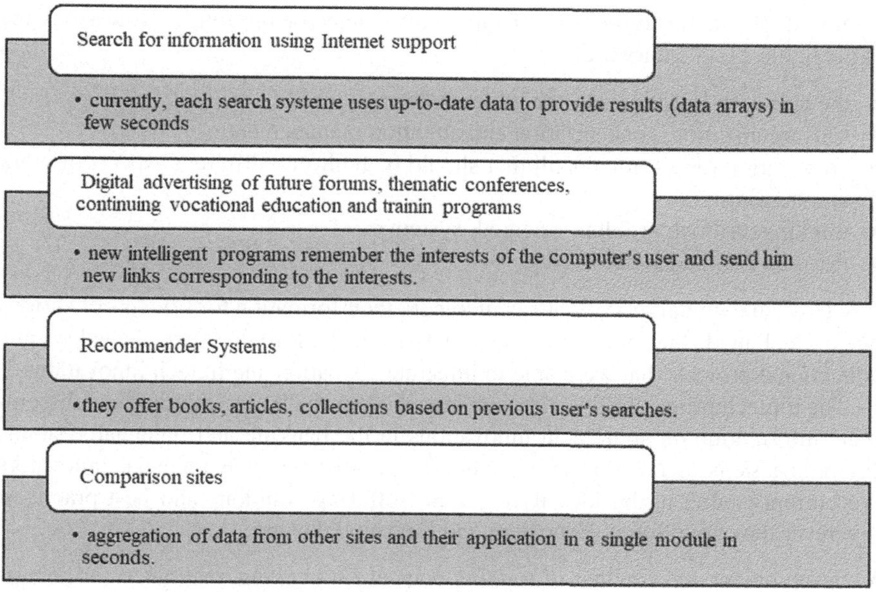

Fig. 3 Application of artificial intelligence in the field of science

Thanks to these gadgets, the emphasis in the formats of the educational process has shifted to such formats as:

- BYOD (Bring Your Own Device). Students are encouraged to use their gadgets in the classroom to search for information, view videos.
- Flipped classroom. Learning the material and completing assignments takes place in the opposite sequence to the standard lesson.
- E-learning. It is not a separate technology, but the creation of a holistic environment that includes learning using mobile devices (mobile learning, or m-learning) and training based on Internet technologies (web-based training, WBT) [14].
- MOOC—(Massive open online courses)—a training course with massive interactive participation. MOOCs help create and support communities of students, teachers, and assistants.

The Future of Education and Skills 2030 aims to help education systems determine the knowledge, skills, attitudes and values students need to thrive in and shape their future. The example of the impact of integration processes in the financial sector on education is the emergence of required future competencies within alternative currency systems (for example, cryptocurrencies such as Bitcoin), reputation economics/merit economics (non-monetary commodity-money exchange, where the person's reputation in the community becomes a currency, including his experience, connections, achievements, etc.), direct investment in talented people (investment in someone else's education and projects on a reimbursable basis) [15].

Based on a detailed analysis of the competencies of the future professions the skills of the 21st century are:

- the ability to learn, unlearn and relearn throughout life;
- self-organization, concentration and attention management;
- cooperation (as a critical skill that should be built into different aspects of work and training);
- thinking: critical, problem-oriented, systemic;
- flexibility and adaptability.

Many educational services are available for their formation for students of various ages, which made the education process massive and affordable [16]. Consider some educational projects that were able to integrate education and fintech innovations.

The Fintech project [17] is a project launched since 2012 to monitor and disseminate information on significant innovations in the banking and financial systems. Its main task is to raise awareness of the opportunities that financial innovators are currently offering by identifying needs, offering solutions and best practices, wherever they are offered. Education and Fintech today are:

- development and conduct of training courses on banking, insurance and investment products;
- gamification of training in the field of finance using examples from the real sector;
- involvement in projects;
- conducting and participating in simulation-business games and championships, including international.

The goal of the School Fundraising Ideas project [18, 22] is to expand the opportunities for students to grow through educational events on ways to publicly raise funds for student initiatives. Project Objectives are:

- achieving the goals of student associations of the university;
- implementation of student association programs;
- strengthening ties between different university student organizations.

The Tutorroom project [19, 23] is a model of the technology market for learning process, it clearly shows how technology is moving from an idea to a finished product and what difficulties the project team may face in the real life. The game has investors, accelerators, employees, technologies at different stages of readiness, the labor market, auction, technology markets and teams. The economy and balance of the game copy the real picture in the technology market.

3 Discussion

In addition to the examples of using modern technologies to support personalized training, it is necessary to separately note the capabilities of the blockchain that may be useful in organizing the learning process:

- storage of students' achieved results/registration and storage of teacher files, which increases their academic mobility, especially relevant for those interested in studying/teaching abroad;
- management and tracking of intellectual assets, since the use of the blockchain creates opportunities not only for tracking assets, but also for managing projects, contracts, such as their insurance through smart contracts.

Our world is becoming more complex, requiring the collective intelligence of many to solve many of the problems we face on a daily basis, most of which not under force to solve one person. First of all, this transformation of reality is reflected in how a modern work that were previously the domain exactly one person, now is a command (project) activities, using their strong hand, resource and collective (batch) ability to achieve the overall objectives.

Educational organizations can continue to work in isolation, fighting for resources, staff, and students. Education in the future must fundamentally reciprocal and collegial at each level.

The second key characteristic of the new format of education became the network form, which is seen nowadays as a tool for increasing the level of competition in High educational system.

To build a network of educational learning communities, you need to understand the principles of the network operating—as a sequence of nodes interconnected by relationships. Such relationships are supported by activity through the channels designed to help the network grow and prosper. It is understood that this requires schools to take into account the focus of their activities (i.e. student learning) and to understand that they are only one of many nodes in a complex series of relationships that, over time, contribute to the learning of the child.

Networking requires the complex work (the first variant is a mutual work of any educational institution and any company, the second variant is a mutual work of two educational institutions) with a specific intent, acceptance of a common goal or purpose and a common way of achieving it. Negative emotional characteristics also present in network cooperation in the form of anticipating criticism of curricula, learning formats by the real business sector. As a rule, network cooperation focuses on improving the educational process. Ruther often some form of critical reflection (formal or informal) supports the development of skills and competencies required and relevant to the real market. Such joint network education can revise the teaching and learning points of view, become effectively, adopt the best practices in different areas of activity including innovation through prototyping, and quickly implement them in their activities. It will avoid the repetition of mistakes and the implementation of ineffective experience.

4 Conclusion

As a result, the list of those areas where modern technologies and artificial intelligence are ready to make big changes in the education system is briefly presented below.

1. Software using artificial intelligence will automate the main activities, for example, certification in the form of testing or solving algorithmic problems.
2. Educational software can be adapted to the needs of students (without classifying them by age, the availability of basic education, etc.).
3. Adaptive programs give the feedback indicated weak or poorly defined places of the courses being studied that can be improved, or students may receive additional advice from relevant teachers.
4. Artificial intelligence is changing the approach of searching and interacting with information (the presence of integrated technologies makes modern students have a completely different experience in researching, searching and processing facts and information, that is, technologies are taught together with students).
5. The shift in the role of the teacher (from the" transmission channel" of data to the "facilitator" of new competencies). Intelligent computer systems help to select the best courses for students and recruitment (matche more clothely the needs and goals of students).

Smart technologies in education also include a mechanism for creating and using virtual content (video conferencing, video lectures), artificial intelligence systems use traditional curricula to create individual textbooks on specific subjects, as a result, textbooks are digitized and new learning interfaces are created to facilitate the educational process for all ages of students. For example, the Netex Learning interface [20] allows to create electronic training programs and training information on many devices. Netex includes online programs, audio, and illustrative videos.

Thus, the general laws of the influence of integration processes, including global ones, on modern learning processes are:

- technology—a tool that provides many people with equal access to skills;
- mass knowledge and skills will be transmitted primarily through automated solutions;
- the learning process is becoming more massive and more total;
- training becomes truly individualized;
- the principle "I can–I know" works.

In general, it can be stated that the goal of modern technologies and technologies based on artificial intelligence is to complement the work of teachers, and not replace them, and reduce their administrative burden so that they can focus on scientific activities or theoretical aspects of their courses. Such measures can also benefit learners.

References

1. Zabrodina I.V., Karataeva T.YU., Kozlova N.A., Sterligova E.A., Fortygina S.N., Fomina L.B., Kuznetsova E.V.: The system of educational situations as a methodological support for updating the content of education. Espacios **40**(9) (2019)
2. Shutenko, E.N., Shutenko, A.I.: Socio-cultural trends in the development of the higher school's innovative potential. Proc. Soc. Behav. Sci. **214**(5), 332–337 (2015). https://doi.org/10.1016/j.sbspro.2015.11.655
3. Zhang, W., Duan, C.: Construction and implementation of institutional e-learning development framework in continuing higher education. J. Educ. Technol. Dev. Exch. **10**(1), 11–25 (2017)
4. Williams, R.D., Smith, C.A.: Constructing the information ground of the campus disability center. Libr. Inf. Sci. Res. **38**(4), 285–291 (2016)
5. Berechet, L.D., Istrimschi, P.A.: Education Reboot: reinventing the University. Proc. Soc. Behav. Sci. **142**, 755–761 (2014). https://doi.org/10.1016/j.sbspro.2014.07.611
6. Stefaniak, J.E.: The implementation of service-learning in graduate instructional design coursework. J. Comput. High. Educ. **27**(1), 2–9 (2015). https://doi.org/10.1007/s12528-015-9092-7
7. Spanjers, I.A.E., Könings, K.D., Leppink, J., Verstegen, D.M.L., de Jong, N., Czabanowska, K., van Merriënboer, J.J.G.: The promised land of blended learning: quizzes as a moderator. Educ. Res. Rev. **15**, 59–74 (2015)
8. Hubackova, S.: Evolution and evaluation of e-learning. Proc. Soc. Behav. Sci. **171**, 231–235 (2015). https://doi.org/10.1016/j.sbspro.2015.01.114
9. Kerimbayev, N.: Virtual learning: possibilities and realization. Educ. Inf. Technol. **21**(6), 1521–1533 (2016). https://doi.org/10.1007/s10639-015-9397-1
10. How Artificial Intelligence enhances education https://thenextweb.com/artificial-intelligence/2017/03/13/how-artificial-intelligence-enhances-education/#.tnw_qirnGnzq
11. Virtual Classroom and Online Learning Management System https://tutorroom.net/edtech/best-examples-educational-technologies/
12. Richard Chang Artificial Intelligence to Grow 47.5% in Education Over Next 4 Years. https://thejournal.com/articles/2017/03/24/ai-market-to-grow-47.5-percent-over-next-four-years.aspx
13. Online Learning Consortium. Quality Scorecard: Criteria for Excellence in Blended Learning Programs. http://onlinelearningconsortium.org/consult/olc-quality-scorecard-blended-learning-programs
14. Al-Rahmi, W., Alias, N., Othman, M.S., Alzahrani, A.I., Alfarraj, O., Saged, A.A., Rahman, N.S.A.: Use of E-Learning by University students in Malaysian higher educational institutions: a case in UniversitiTeknologi Malaysia. IEEE Access **6** (2018). https://doi.org/10.1109/ACCESS.2018.2802325
15. OECD Future of Education and Skills 2030 https://www.oecd.org/education/2030-project/
16. E-Learning Industry. 6 Blended Learning Models: When Blended Learning Is What's Up For Successful Students. https://elearningindustry.com/6-blended-learning-models-blended-learning-successful-students
17. Fintech Association. https://fintechru.org/
18. Proven School Fundraising Ideas—Raise Money for Your School. https://donorbox.org/nonprofit-blog/school-fundraising-ideas/
19. Tutorroom. Online Teaching Software. Best examples of Educational Technologies. https://tutorroom.net/edtech/best-examples-educational-technologies/
20. Netex Knowledge Factory 2019. https://www.netexlearning.com/en/about-us/
21. Natocheeva N., Borodin A., Rud N., Kutsuri G., Zholamanova M, Namitulina N.: Development of tools for realizing the potential of financial stability of enterprises. Entrep. Sustain. Issues **7**(2), 1654–1565 (2019). https://doi.org/10.9770/jesi.2019.7.2(60)

22. Shash N., Panaedova G. , Frumina S. , Kairbekuly A., Mityushina I.: The impact of the publication of non-financial statements on the financial performance of companies with the identification of intersectoral features. Entrep. Sustain. Issues **7**(2), 1666–1685 (2019). https://doi.org/10.9770/jesi.2019.7.2(61)
23. Ziyadin S., Borodin A., Streltsova E., Suieubayeva S., Pshembayeva D.: Fuzzy logic approach in the modeling of sustainable tourism development management // Pol. J. Manag. Stud. 1(19), 492–504 (2019). https://doi.org/10.17512/PJMS.2019.19.1.37

Sustainable Development in the context of Digitalization: Measures that Drives Performance

Digital Technologies in the Oil and Gas Industry: Main Directions and Development Trends

Galina Panaedova, Svetlana Frumina, and Irina Mityushina

Abstract Today, digitalization, in particular working with big data, is highly popular and deeply embedded in all economic processes. In the context of our time, it is considered digital technologies that improve business processes in the field of automation, computer computing, and data digitalization. Using these technologies, it is possible to accurately design, manage, control and change processes in the virtual space, which can be carried out with greater speed and efficiency. The article deals with the issues of digitalization of the oil and gas industry. The main levels of the digital economy are given. Statistical data on the business digitalization index by country is provided. The focus is on the revolutionary trends in the oil and gas industry. The "industry 4.0" project for oil and gas companies is presented, which reflects the concepts of smart wells and fields. The analysis of projects on digitalization of oil companies that have been implemented and are planning to be implemented in the near future, as well as the companies studied the level of prospects for the use of digital (smart) technologies. The rating of the most expensive oil and gas companies in terms of capitalization for 2013–2018 is presented. The analysis of data with the use of digital technologies in companies is made. The main issues of digital products that are not widely used in the oil and gas sector are considered. The practical application of digital products in the oil and gas sector of the economy is analyzed. The main results and prospects of digitalization of the oil production complex business are revealed.

Keywords Digital economy · Smart wells · Internet of things · Smart fields · Digital doubles · Big data · Economic transformation · Internet technologies in the oil and gas industry

---

G. Panaedova (✉)
North-Caucasus Federal University, 1, Pushkin str, Stavropol, Russian Federation
e-mail: afina-02@rambler.ru

S. Frumina
Financial University Under the Government of Russian Federation, 49, Leningradskij prospekt, Moscow, Russian Federation

I. Mityushina
Belgorod National Research University, 85, Pobedy Street, Belgorod 308015, Russian Federation

G. Mutanov and A. Serikbekuly (eds.), *Digital Transformation in Sustainable Value Chains and Innovative Infrastructures*, Studies in Systems, Decision and Control 443,
https://doi.org/10.1007/978-3-031-07067-9_6

1 Introduction

The term digital economy was first introduced by Nicholas Negroponte in 1995, a well-known scientist who created the media laboratory at MIT (the Massachusetts Institute of Technology is one of the most prestigious universities in the field of technology in the world).

Another interpretation of the digital economy implies that it is an extended concept of the Internet economy, which includes related market segments [1].

This understanding has been revised in recent years due to the fact that many industries, including those classified as "real sector", are undergoing a digital transformation.

In Russia, the approved form of The "strategy for the development of the information society of the Russian Federation for 2017–2030" mentions that the concept of Digital economy is an economic activity in which the key factor of production is data in digital form, processing large volumes and using the results of analysis of which, in comparison with traditional forms of management, can significantly increase the efficiency of various types of production, technologies, equipment, storage, sale, delivery of goods and services [2–4].

The digital economy can be represented as a three-level system [5].

The first system includes an environment in which conditions are created for the development of high-performance technologies.

The second system is presented as a technology and platform.

The third reflects economic sectors and markets where service providers and consumers interact directly [6].

The digital economy should be identified as a priority of knowledge. At the same time, according to the international consulting company McKinsey, there is no research on the opportunities and potential of the digital economy in Russia [7]. However, in Russia, we can observe the growth of scientific research on the use of digital economy and financial technologies, as well as the impact of these trends on social and economic relations, and the modernization of public institutions. Today, global projects are being created to digitalize and transform the economy.

2 Results

The growth of digital technologies, digitized data, and the use of virtual space lead to the transformation of companies' business activities [8, 9]. The relevance of digitalization has reached a high level and entered all spheres of the economy. Active implementation of digital products is taking place in all countries of the world.

The leader for 2017–2018 remains the country of Finland with 50 points (Fig. 1). Russia ranks 26th in terms of digital technology adoption, having improved the indicator by 3 points in 2018 (the indicator is 31 points in 2018) [10, 11].

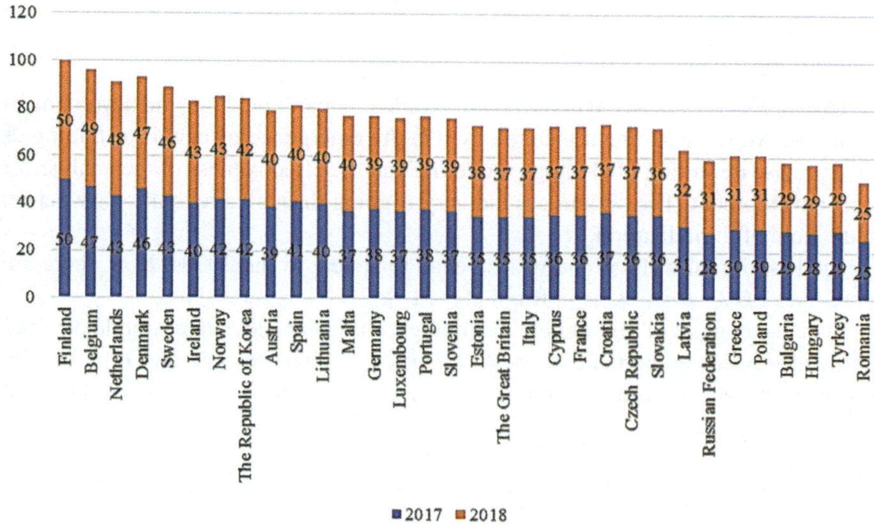

Fig. 1 Business digitalization index by country for 2017–2018, the measurement is presented in points

On the basis of a new stage of transformation at the state level, the program "Digital economy of the Russian Federation" has been developed, where the main reference point is Russia's entry into the number of global leaders in digitalization by 2024 [12, 13]. The implementation of the program provides for the following issues:

(1) creating an information structure;
(2) introduction of digital technologies in all areas of activity;
(3) training of digital personnel;
(4) creation of a high-quality regulatory framework. One of the priority areas of the economy is the oil and gas industry, which provides more than 30% of the funds to the state budget.

The oil and gas production complex includes almost the entire list of digital products (Fig. 2). In terms of potential benefits, the greatest prospects are big data, the Internet of things, and digital counterparts, which are called one product "Smart" field.

Despite the fact that Russian companies do not occupy the first places in the ranking of the most expensive companies in the world, they are actively implementing digital technologies in their business and actively using Industry 4.0 technologies. Key technologies of "Industry 4.0" include:

- Big data;
- Internet of things;
- Robots;
- Digital doubles;
- Smart materials;

- 3D printing;
- Distributed registry (Fig. 2).

Two factors played a role in the creation of the "Smart" field: (1) the development of remote well management technology; (2) the emergence of new it technologies put a precise definition of Geology modeling and the effectiveness of the analysis of asset development. The oil and gas industry is going through a revolutionary period in the field of digitalization.

The first digital wells in the world began to appear in the 2000s. Now, there are more than 2,000 smart wells in Russia [14, 15].

Also, the content of digital field technologies has changed, smart mobile devices, digitized modern equipment, drones, robotics for obtaining fast and high-quality information and substitutes for live labor for routine work, as well as the most

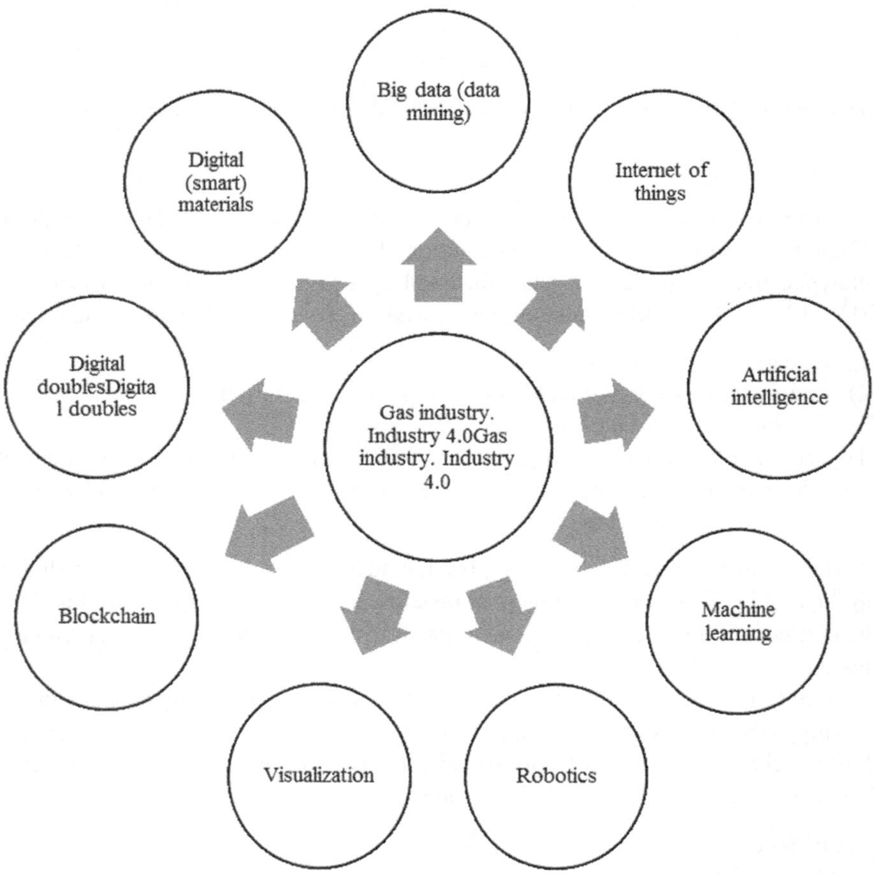

Fig. 2 Application of Industry 4.0 in the oil and gas sector Application of Industry 4.0 in the oil and gas sector. *Note*: compiled by the author based on data from VYGON Consulting [16]

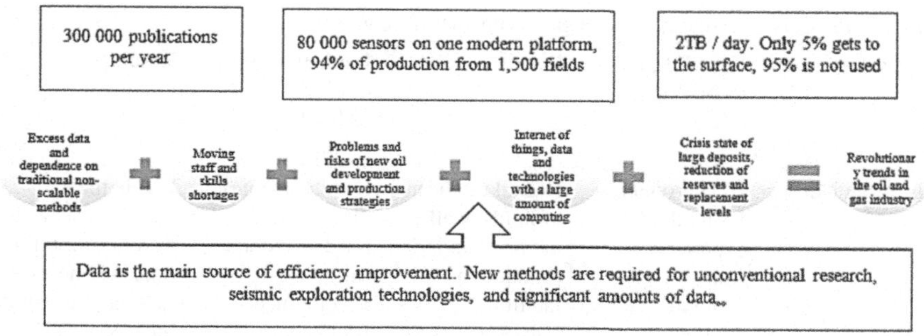

Fig. 3 Revolutionary trends in oil and gas production have grown. *Note*: compiled by the author based on https//www.ibm.com

important appearance of robots for analyzing and managing oil production in real time (Fig. 3).

Gazprom Neft is one of the leaders in the use of digital technologies and uses them in all aspects of its operations (from field development to well operation and maintenance). PJSC Gazprom Neft has created a "Unified management", which consists of remote control of the construction of several dozen wells at once. The use of artificial intelligence leads to rapid analysis of calculations, models and combinations of equipment installation. Advantages of artificial intelligence: decision-making efficiency increases by 20–25% and reduces the time of finding drilling sites from several months to several minutes.

The advantages of the digitalization of the PJSC "Tatneft" are as follows:

developed a project for digitalization, the result of which led to a reduction in the cost of production by 30%;
the use of digital doubles and models allows you to choose the best places for drilling and placement of wells;
mobile technologies—building a corporate network.

PJSC LUKOIL has developed the Digital LUKOIL 4.0 Project in 4 areas:

(1) Digital doubles;
(2) Optimizing staff performance;
(3) Robotics;
(4) Digital ecosystem.

For more detailed information on companies, see Table 1.

In fact, we are already seeing progress in the activities of the companies using digital technologies. This is shown in terms of capitalization (Table 2). The change in this indicator in 2018 is significant; Rosneft exceeded it by 22.5% compared to 2017. LUKOIL's capitalization increased by 9.9% in 2018 compared to 2017. During this time period, NOVATEK (in 2018 by 38.97% compared to 2017) and Tatneft (by 27.28%) also improved their performance (Table 2).

Table 1 Projects to digitalize oil and gas companies have grown

Company	Use of digital technologies (fact)	Result	Implementation of digital technologies (plan)	Result
Gazprom Neft	IBM product; Using the GE experience; Using "Digital doubles»; Digital materials (equipment) Cyber security	Digital duplicates of fields, wells, factories and production facilities Use of the latest technology(equipment), gadgets, automation of all processes of the company Data security and developed solutions and strategies	Development of a digital platform for data storage (in-house development); Blockchain Internet of things Machine learning algorithms and artificial intelligence	Use in their activities, as well as replication for other segments (players) of the business Speed of interaction of the entire chain of the activity process Control of oil production and refining processes For storing large amounts of data
Lukoil	The program "Digital LUKOIL 4.0" was developed»	(1)Digital doubles—a set of models for describing an object in full, not in parts, or in separate details (2)Optimization of staff work (3)Robotics is the use of software robots that can relieve employees engaged in routine work	(1)It is Planned to build an ecosystem that includes a single digital workspace (2)Intellectual field	Field management is based on automated computer systems and high-tech data collection systems

(continued)

3 Discussion

Today, Russia has all the prospects for further development of the digital economy. New technologies of digitalization processes will affect not only business as a whole, but also public administration and the growth of the quality of life of the population.

An important aspect is high-quality digital education and the growth of professionalism of employees who could perfectly control and work with the use of digital technology.

Table 1 (continued)

Company	Use of digital technologies (fact)	Result	Implementation of digital technologies (plan)	Result
Rosneft	The pilot project "Digital oilfield" with the company Bashneft	Main elements; digital counterparts, the latest mobile devices with data transmission, digital (intelligent) pipeline monitoring system. Use of drones with thermal imaging control and video Analytics system	Integrated digital solutions project	(1)"Digital field»; (2)"Digital factory»; (3)"Digital supply chain»; (4)"Digital gas station"
PJSC "Bashneft»	Launched the Digital field project	Main elements; digital counterparts, the latest mobile devices with data transmission, digital (intelligent) pipeline monitoring system. Use of drones with thermal imaging control and video Analytics system	The project "Digital twins»	Development of digital models of technical models. Improving data cybersecurity and employee security
PJSC Tatneft	(1)Creating Big Data and converting it to Smart Data; (2)Coverage and instant access to all information resources of the company; (3)Creating a system to reduce the time of information provided; (4)Creating digital platforms	(1)Reduced data preparation time by 30–60%; (2)Improving data quality; (3)Optimized the data storage system; (4)The Ability to organize complex Analytics based on big data analysis technology	The project "Digital transformation of business»	By 2025, the capitalization will increase by 2 times, and the production volume will increase by 11%

Note: Compiled by the author on the basis of official data [17–22]

Table 2 The largest Russian oil and gas companies in terms of capitalization for 2013–2018, million dollars

Company	2013	2014	2015	2016	2017	2018	Change in 2018–2013 (%)	Change in 2018 г.–2017 г., (%)
Rosneft	87,266	47,136	37,258	69,907	53,304	65,286	−25,2	22,5
Lukoil	52,128	30,658	27,457	48,076	48,993	53,823	3,2	9,9
Gazprom Neft	104,386	58,451	44,258	59,932	53,349	52,240	−49,95	−2,1
NOVATEK	32,063	21,669	24,806	39,220	35,543	49,393	54,05	38,97
PJSC Tatneft	14,488	8023	9548	15,123	17,959	22,859	57,78	27,28
Surgutneftegaz	29,288	17,957	16,745	18,217	17,191	13,808	−52,85	−19,7
PJSC "Bashneft»	10,176	6692	4030	8754	5793	3984	−60,8	−31,22

Compiled by the author on the basis of Rosstat data [14]

Another factor in the development of digital technologies may be non-financial reporting of oil and gas companies. Depending on which effect is stronger, disclosure of non-financial information can have both a positive and negative impact on financial performance [23, 24]. Non-financial reporting may include a "Smart ecology" section, in the form of digital treatment facilities.

The oil and gas sector is currently focusing on digital doubles, but blockchain is considered a very important product of digitalization. In 2018, Gazprom (one of the Russian companies) launched a pilot project using blockchain, which provides for solving financial, trade and logistics tasks. Therefore, this product can completely transform the oil and gas industry.

4 Conclusion

- Digital technologies are gaining momentum and are rapidly being implemented in all areas of the economy;
- the main areas of digitalization of "Industry 4.0" are as follows: robots, the Internet of things, smart materials, 3D printing, big data (machine learning, data mining, artificial intelligence, visualization, digital doubles, distribution of registers;
- the oil and gas Industry is one of the most promising sectors for digital projects;
- developed projects based on companies lead to an increase in business capitalization and profitability;
- in the roadmaps "Regulatory regulation" and "Technological groundwork", you can find a sufficient number of measures and proposals to accelerate the development of digital transformation of Russian oil production by 2024.

References

1. Averyanov, D.A., Stages of development of the digital economy. In: Averyanova, D.A. (ed.) Economy and Bisiness, vol. 1, pp. 10–13 (2019)
2. Order of the government of the Russian Federation No. 1632-R "program" Digital economy of the Russian Federation". http://static.government.ru/media/files/9gFM4FHj4PsB79I5v7yLVuPgu4bvR7M0.pdf (2017). Accessed 25 Jan 2020
3. Decree of the President of the Russian Federation No. 203. On The strategy for the development of the information society in the Russian Federation for 2017–2030. http://kremlin.ru/acts/bank/41919 (2017). Accessed 25 Jan 2020
4. Order of the Government of the Russian Federation "on approval of the program" Digital economy of the Russian Federation" No. 1632-R—Mode of access: Reference legal system "Consultant Plus" (2017). Accessed 25 Jan 2020
5. Borodin, A.I., Vaganova, O., Vladyka, M., Kucheryavenko, S.: Investments as an accelerator of regional innovation development. Int. Bus. Manag. **10**(16), 3438–3442 (2016)
6. Mityushina, I.L.: Stages of digital economy development in Russia and in the world. In: Belgorod, Ed. (ed.) Economic and management Congress: collection of articles based on the materials of the Russian scientific and practical event with international participation of NRU "BelSU", p. 380C. Belgu publishing house (2019)
7. Dunenkova, E.N.: The development of innovation in the digital economy. In: Dunenkova, E.N., Gureev, P.M., Prokhorova, I.S., Boldyrev, V.A., Kamchatova, E.Yu., Faustov, A.A., Degtyareva, V.V (eds.) pp. 19–33. State University of management, Moscow (2019)
8. Strategy in the oil and gas sector. https://www.dohod.ru/analytic/strategiya-v-neftegazovom-sektore-4-kvartal-2019-goda/. Accessed 27 Jan 2020
9. Kostyukov, K.I., Svechinskaya, T.A.: "Digital economy" as a factor of modern development of the Russian economy. In: Kostyukov, K.I., Svechinskaya, T.A.(eds.) In the collection: Geopolitics and geo-Economics: problems and solutions Collection of scientific papers of the International scientific and practical conference (Edited by K.I. Kostyukov), pp. 33–38 (2018)
10. Data of Russia for oil production. https://www.gazeta.ru/business/2018/12/18/12099061.shtml. Accessed 10 Dec 2019
11. The Ministry of energy, mining, petroleum based. https://minenergo.gov.ru/node/1209. Accessed 27 Jan 2020
12. Digital transformation of the oil and gas industry Modernization of development and production using innovative technologies. https://www.ibm.com/downloads/cas/JLE286ZX. Accessed 27 Jan 2020
13. Borodin, A.I., Ziyadin, S., Doszhan, R., Omarova, A., Ilyas, A.: The role of social media marketing in consumer behavior. E3S Web of Conferences **135**, 04022 (2019). https://doi.org/10.1051/e3sconf/201913504022
14. Rating of the most expensive public companies in Russia by capitalization. https://riarating.ru/infografika/20180130/630080911.html. Accessed 26 Jan 2020
15. Eremin, A.N., New classification of digital and intelligent wells. In: Automation and IT in the oil and gas field, vol. 2, 24th edn, pp. 20–22 (2016)
16. Kozlova, D., Pigarev, D.: Digital oil production: tuning for growth. https: vygon.consulting/upload/iblock/d11/vygon_consulting_digital_upstream. Accessed 27 Jan 2020
17. Official website of PJSC Gazprom Neft. https://www.gazprom-neft.ru/press-center/news/v-2015-godu-gazprom-neft-uvelichila-zapasy-uglevodorodov-na-5-6-vozmestiv-dobychu-na-279/. Accessed 26 Jan 2020
18. Official website of PJSC LUKOIL. http://www.lukoil.ru/. Accessed 26 Jan 2020
19. Official website of PJSC Tatneft. https://www.tatneft.ru/aktsioneram-i-investoram/raskritie-informatsii/godovie-otcheti?lang=ru. Accessed 26 Jan 2020
20. Official website of PJSC "Rosneft". https://www.rosneft.ru/press/news/item/198039/. Accessed 26 Jan 2020

21. Monthly Information and analytical publication. Joint project of the oil and gas forum and exhibition "Neftegaz". https://www.neftegaz-expo.ru/common/img/uploaded/exhibitions/neftegaz/doc_2019/Neftegaz_Digest_2019.10(17).pdf. Accessed 26 Jan 2020
22. RBC. Market research. http://research.rbc.ru. Accessed 27 Oct 2019
23. Borodin, A., Shash, N., Panaedova, G., Frumina, S., Kairbekuly, A., Mityushina, I.: The impact of the publication of non-financial statements on the financial performance of companies with the identification of intersectoral features. Entrep. Sustain. Issues. **7**(2), 1666–1685 (2019). https://doi.org/10.9770/jesi.2019.7.2(61)
24. Abukova, L.A. et al.: Digital modernization of the oil and gas complex of Russia. In: Abukova, L.A., Dmitrievsky, A.N., Eremin, N.A. (eds.) Oil economy, vol. 10, pp. 54–58 (2017)

The Effective Marketing Methods and Their Use in the Field of Ecology

Zauresh Akhmetova, Bakytgul Rysbayeva, Assel Kozhakhmetova, and Nadira Turlybekova

Abstract In this article, we will look at the main marketing approaches in the ecology field. The study examines the influence of marketing on ecology. Aspects of environmentally oriented marketing in the global community are associated with the rapid development of technologies and processes that reduce the environment's impact. Formation of the market for environmental services, of course, requires the appropriate development of marketing management tools. Research findings help to consider the impact of factors and results of environmental impact on marketing strategies. As a result, the study develops a conceptual framework consisting of three sets of designs. The first point focuses on the firm's resources. The next one is green marketing policy focused on creating a competitive advantage. The third set shows the factors of the competitive advantage. Organizational capabilities and strategies are considered, SWOT analysis and PEST analysis are made.

Keywords Ecology · Environmental marketing · Marketing approaches · Efficiency · Green marketing · Development · Eco-labeling · Ecosystem

1 Introduction

With the intensification of the environmental problem on the planet, various groups of stakeholders (consumers, regulators, the General public) are putting increasing pressure on firms to take decisive measures to protect and maintain the natural environment. This filed is relatively new for marketing. One of the issues that deserves special attention is the special role of companies' resources used for the preparation of marketing strategies, in addition, impact on business performance. While previous studies have looked at firms' environmental practices as a resource or capability that provides a competitive advantage, special attention focuses on the role of resources and capabilities that are responsible for shaping environmental

Z. Akhmetova (✉) · B. Rysbayeva · A. Kozhakhmetova · N. Turlybekova
Al-Farabi Kazakh National University, al-Farabi Ave. 71, 050040 Almaty, Republic of Kazakhstan
e-mail: Axmetova.Zauresh@kaznu.edu.kz

G. Mutanov and A. Serikbekuly (eds.), *Digital Transformation in Sustainable Value Chains and Innovative Infrastructures*, Studies in Systems, Decision and Control 443,
https://doi.org/10.1007/978-3-031-07067-9_7

clean marketing and for the results of its implementation. Because research on environmental marketing or management has mostly focused on manufacturing firms, because of their greater impact on environmental issues in the form of costs and benefits, the research should also extend to "green" issues in the service sector, which is often described as environmental destruction [1].

Current thoughts about an environment are seen in consumer surveys and lifestyles, in political campaigns, in the media, as well as in product offerings and marketing strategies. However, academic literature has not yet reflected this movement. Our research identifies "green" issues—those aspects of the environment that are currently considered important by consumers. Then we discuss the relationship between consumers' perception of green issues and marketing strategies. Research findings are valuable for investigating consumer behavior and marketing researchers, marketing managers, and policy makers at a time when consumers are increasingly aware of environmental issues.

2 Literature Review

Currently, the literature on trade friction mainly analyzes social and economic effects such as trade flow, welfare, etc.—losses during periods, and economic losses This research will instead focus on the environment effects. Commodity imports and exports also mean the transfer of emissions between countries. These trade-embodied environmental emissions not only contribute significantly to global emissions, but also significantly affect regional distribution of emissions: empirical results show that uneven exchange in the environment stills sufficient despite of low rates of domestic pollution.

Ecological marketing is a new business philosophy that organizations can use to solve sustainable development problems. They analyze and propose to use a model that analyzes how the implementation of ecological issues within the framework of the marketing strategy and the direction of the firm's work will affect organizational results [2].

The authors Keenan et al. presented and analyzed the results of the 2015 global forest resources assessment [3].

The work of scientists Mutanov et al. was aimed at developing a graphical model focused on innovation evaluation, competitiveness, and environmental effectiveness of eco-innovative projects. When developing this method, specialists used a methodological approach [4].

The authors Vinod and Zilur in their work consider the concepts of marketing strategy and sustainability, discuss the evolution of sustainability in marketing strategy [5].

3 Methodology

The marketing mechanism of environmental management is based on the typology of market methods. Currently, the following main groups of management methods are known administrative regulation-the introduction of appropriate regulatory standards and restrictions that manufacturers must comply with; direct control and licensing of environmental management processes; economic incentives aimed at attracting the interest of the manufacturer in environmental management system of payments for pollution; distribution of pollution rights and compensation payments [6]. SWOT analysis, systematic literature review, model building, classification, deduction methods were used for conducting this research.

4 Results

Corporations are struggling to market products that are positioned and advertised as meeting the needs of an environmentally conscious consumer. However, the vast majority of claims submitted in support of these products are best described as confusing, misleading, or completely illegal. Ethical considerations have not yet been incorporated into environmental marketing, and as a result, there may be long-term damage at both the individual and societal level. The sequence is based on 2 approaches of ethic theory: normative behavior and moral style [7].

Recent research has called for studies connecting performance and environmental strategies. This study sets in motion the environmental marketing (EM) construct and explores its relationship to the firm's performance. This is the first empirical study for the practical implementation of the EM construction [8].

EM directly affects the capabilities of the company (for example, the success of new product development), but not its advantage (changes in market share). A nationwide survey of top-level marketing managers supports this view. In addition, although market turbulence also affects the success of new product development, it does not affect EM. This suggests that the formation of EM is due to internal, not external forces [9].

Considering the impact of factors and results of environmental impact of strategies, the study prepares the framework including 3 sets of structure. Firm's resources and capabilities that work as driving forces of strategy [10].

The next one is green marketing policy focused on creating a competitive advantage. The relationship between these two constructions is assumed to be moderated by both the intensity of competition and the dynamism of the market. The third set shows that an environmentally sound competitive advantage leads to higher market and financial performance [11].

Our model is based on a resource-based view (RBV) that emphasizes the firm's resources as key competitive factors advantages and effectiveness of the business. Then the firm gets a competitive advantage by building strategies.

(1) organizational resources are an important contribution to the development and implementation of effective environmental marketing strategies. The study found 3 groups: resources in the relevant literature that play a crucial role in the development of ecologically friendly marketing strategies. The first is physical resources, which are understood as the possession of modern technologies, equipment and the availability of production capacity or other infrastructure necessary to improve the efficiency and efficiency of the product/service. When used correctly, these resources help the firm make better use of internal resource methods related to ecological issues such as waste reduction, energy conservation, and Ecological Pollution Prevention [12].

H1. The more physical resources a firm has, the more it can develop a marketing strategy for the environment.

The second type of resources are financial dimensions, such as the firm's financial liquidity, working capital, and leverage, which are crucial in achieving stability and success in green marketing. Investing in clean technologies and green practices involves:

(1) significant amounts of expenditure (especially in the initial stage); the new technology stage that take a long time to achieve a satisfactory profit margin; (2) high risks associated with the development and support of green strategies, especially in fast-growing industries such as the hotel sector; and (3) Indeed, restrictions on financial resources constitute one of the main barriers that hotels face, especially small ones, to taking environmental measures (Fig. 1).

H2. The more financial resources a firm has, the greater the opportunity to develop an environmental marketing strategy.

The third resource has an empirical nature, i.e., knowledge obtained from the firm's experience, which helps to identify and compare customer needs and anticipate new market trends. With the adoption of internal procedures and the accumulation of knowledge related to environment, the firm expands its expertise knowledge that acts as an intermediary in building environmentally friendly marketing strategies.

Organizational capabilities and strategy

The literature on environmental marketing highlights several features that accompany an environmental marketing strategy. Firms that develop these capabilities will be able to implement sustainable business practices, develop an environmentally sensitive culture, better understand the requirements of various stakeholders, and develop sound marketing strategies and processes around them. We have tracked three types of opportunities that affect an eco-friendly marketing strategy. The first ability is a shared vision, which consists of having shared ideas in achieving green organizational goals. Firms with a shared vision are able to collect and organize the resources needed for sustainable development. As an example, we can quote the words of Hilton: "we care!" the program (involving 16,000 employees), in which the creation of

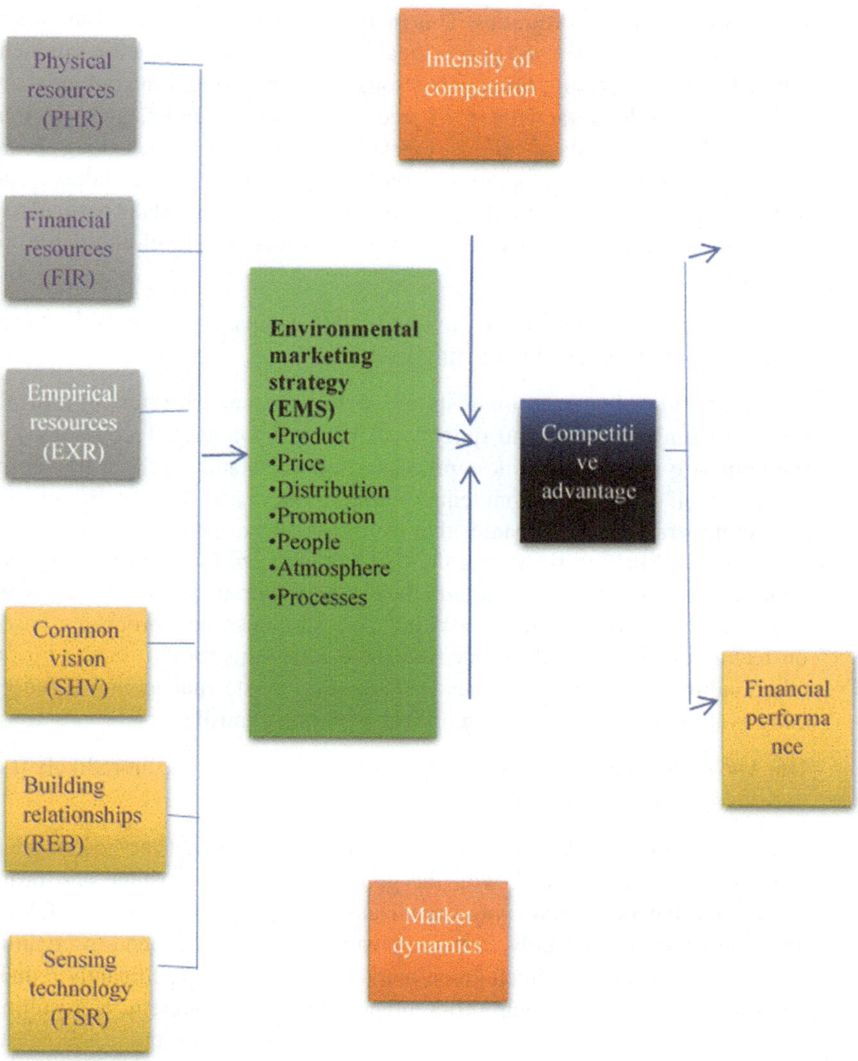

Fig. 1 Conceptual model

special initiative groups linking all levels of employees in the organization, was responsible for significantly reducing energy, water and carbon dioxide emissions [13].

H4. The firm that has more opportunities more open for develop an environmental marketing strategy.

By building relationships, firms gain a better understanding and thus can better respond to their needs and desires, as well as identify potentially profitable market trends and opportunities. This increases the sensitivity of stakeholders to environmental issues and is a critical force in making firms environmentally friendly. As a result, firms with strong relationship-building potential are better aware of the attractiveness of green segments and understand the environmental requirements of various stakeholders in specific markets long before competition begins. Developing relationships with various stakeholders (such as tour operators) and responding to their environmental needs is particularly important in the hotel business, as it directly affects the level and nature of demand.

H5. The higher the firm's ability to build relationships, the more likely it is to develop an environmental marketing strategy.

The third ability is perception/response technology, which refers to a firm's ability to sense and respond quickly to new technologies. Technology has the ability to influence business processes, products and services, transform them, as well as adapt environmental settings to them and form environmental marketing strategies. Companies that can sense and respond to technological advances are likely to be among the first to recognize the benefits of implementing green technologies, identify the most appropriate and least risky clean technologies to implement, and build their strategies and processes based on technologies that will produce the best economic results. In the hospitality industry, such technologies include solid waste management, energy conservation, water conservation and air pollution control.

H6. The higher the firm's ability to perceive and respond, the more likely it is to develop an environmental marketing strategy.

an environmental marketing Strategy includes activities that include an environmentally friendly focus, with the goal of generating revenue and profit while simultaneously achieving organizational and individual goals. Taking an environmentally friendly strategic position in hotels can lead to creating competitive advantages. Such green marketing strategies significantly reduce costs in the long term or help differentiate offers from competitors, resulting from the use of cheaper secondary materials, energy-saving processes, waste minimization solutions, and workflow improvements. An example is the Hyatt Regency Chicago, which, thanks to its special recycling program, recovered about 70% of used products (such as towels, dishes, and bed linen). The ability to focus on an environmentally friendly consumer segment also contributes to competitive advantage. All these advantages offer customers a more attractive, attractive, profitable, and acceptable company than their competitors (Tables 1 and 2).

H7. Environmental marketing strategy adoption is associated with achieving a competitive advantage [14].

Table 1 SWOT-analysis

Weakness * Most customers prefer to satisfy their personal needs before taking care of the environment * Excessive emphasis on greenery rather than customer needs can be disruptive to the product	**Strengths** * Marketers can perceive products positively if they looks like environmentally responsible * The company that implement green marketing will be perceived as more eco responsible * Green marketing helps to build brand's awareness and loyalty between company and consumers
Opportunities * Marketing in a market that became more environmentally aware and responsible. Consumers demand products that match this new attitude * Organizations perceive green marketing as a competitive advantage over their competitors	**Problems** * Uncertainty about the environmental impact of any activity, including those that are perceived as less environmentally harmful * Uncertainty about which green marketing activities are acceptable from the government's point of view

Table 2 PEST analysis

P (Political) Introduction of a law that will limit the possibility of extra charges on goods; changes in taxation in the industry or the introduction of new tariffs; introduction of a law that restricts advertising or distribution of goods; stricter requirements for product certification	**E (Economical)** Next year, the forecast for GDP growth is 3% The availability of loans will increase Inflation is forecast at 7%, and unemployment is expected to fall to 10%, The level of disposable income of the population will decrease
S (Socio-cultural) There is a growing positive attitude towards natural products in society The number of Internet purchases and mobile Internet usage in the age category under 35 is increasing The fear of lending is reduced People prefer to spend their income and are not set up to form long-term savings	**T (Technological)** There is no technological rush in the industry The market values design more than technological properties of the product Research and new developments are conducted only in the field of improving the sensor In flavorings, a technology is being developed that allows you to reduce the cost of raw materials by 15%

5 Conclusion

Our analysis shows that RBV can provide a reliable theoretical platform for explaining the prerequisites and results of adopting an eco-friendly marketing strategy. In addition, this advantage will be very helpful for companies operating in quite competitive environment since it helps distinguishes them from other competitors. In turn, a competitive advantage based on the environment should allow firms to achieve higher market and financial performance [15].

Having sufficient resources (especially physical and financial) contributes to the establishment of a rational environmental marketing strategy and help it maintain favorable marketing programs. However, to effectively influence on nature friendly sphere these findings should be unique [16, 17].

Most of the environmental results affected by regional difference. It's not just about taxes, payments for natural resources, for emissions and discharges of pollutants and waste disposal, about the various environmental benefits and sanctions, but also on state and public ecological expertise, which are pre-planned documentation, justification, feasibility study, proposals, specifications, projects, production objects, their anthropogenic impact on the environment, technology, equipment, products, and wastes [18–21].

References

1. Baker, W.E., Sinkula, J.M.: Environmental marketing strategy and firm performance: effects on new product performance and market share. J. Acad. Mark. Sci. **33**(4), 461–475 (2005)
2. Davis, J.J.: Ethics and environmental marketing. J. Bus. Ethics **11**(2), 81–87 (1992)
3. Fraj-Andrés, E., Martinez-Salinas, E., Matute-Vallejo, J.: A multidimensional approach to the influence of environmental marketing and orientation on the firm's organizational performance. J. Bus. Ethics **88**(2), 263 (2009)
4. Keenan, R.J., Reams, G.A., Achard, F., de Freitas, J.V., Grainger, A., Lindquist, E.: Dynamics of global forest area: results from the FAO Global Forest Resources Assessment. For. Ecol. Manag. **352**, 9–20 (2015)
5. Mutanov, G., Ziyadin, S., Shaikh, A.: Graphic model for evaluating the competitiveness and eco-efficiency of eco-innovative projects. Entrep. Sustain. Issues **6**(4), 2136–2158 (2019). https://doi.org/10.9770/jesi.2019.6.4(41)
6. Kumar, V., Rahman, Z., Kazmi, A.A., Goyal, P.: Evolution of sustainability as marketing strategy: beginning of new era. Proc. Soc. Behav. Sci. **37**, 482–489 (2012)
7. Murphy, P.E., Laczniak, G.R.: Marketing and ecology: retrospect and prospect. Bus. Soc. **18**(1), 26–34 (1977)
8. Khalid, S., Shahid, M., Niazi, N.K., Murtaza, B., Bibi, I., Dumat, C.: A comparison of technologies for remediation of heavy metal contaminated soils. J. Geochem. Explor. **182**, 247–268 (2017)
9. Liu, L.J., Creutzig, F., Yao, Y.F., Wei, Y.M., Liang, Q.M.: Environmental and economic impacts of trade barriers. Resour. Energy Econ. **59**, 101144 (2020)
10. .Melo-Merino, S. M., Reyes-Bonilla, H., Lira-Noriega, A.: Ecological niche models and species distribution models in marine environments: a literature review and spatial analysis of evidence. Ecol. Model. **415**, 108837 (2020)
11. McKay, S.K., Schramski, J.R., Conyngham, J.N., Fischenich, J.C.: Assessing upstream fish passage connectivity with network analysis. Ecol. Appl. **23**(6), 1396–1409 (2013)
12. Ziyadin, S., Doszhan, R., Borodin, A., Omarova, A., Ilyas, A.: The role of social media marketing in consumer behaviour. E3S Web of Conferences, vol. 135, p. 04022. EDP Sciences (2019)
13. Pärssinen, M., Kotila, M., Cuevas, R., Phansalkar, A., Manner, J.: Environmental impact assessment of online advertising. Environ. Impact Assess. Rev. **73**, 177–200 (2018)
14. Polonsky, M.J.: A stakeholder theory approach to designing environmental marketing strategy. J. Bus. Ind. Mark. **10**(3), 29–46 (1995)
15. Vidal-Abarca, M.R., Santos-Martín, F., Martín-López, B., Sánchez-Montoya, M.M., Alonso, M.S.: Exploring the capacity of water framework directive indices to assess ecosystem services in fluvial and riparian systems: towards a second implementation phase. Environ. Manage. **57**(6), 1139–1152 (2016)

16. Leonidou, L.C., Leonidou, C.N., Fotiadis, T.A., Zeriti, A.: Resources and capabilities as drivers of hotel environmental marketing strategy: implications for competitive advantage and performance. Tour. Manage. **35**, 94–110 (2013)
17. Hahladakis, J.N., Velis, C.A., Weber, R., Iacovidou, E., Purnell, P.: An overview of chemical additives present in plastics: migration, release, fate and environmental impact during their use, disposal and recycling. J. Hazard. Mater. **344**, 179–199 (2018)
18. Saparbayev, A., Demesinova, A.: Development strategy for the energy sector in a region with scarce power resources: approaches and principles. Actual Probl. Econ. **152**(2), 262–272 (2014)
19. Zimmer, M.R., Stafford, T.F., Stafford, M.R.: Green issues: dimensions of environmental concern. J. Bus. Res. **30**(1), 63–74 (1994)
20. Mutanov, G., Ziyadin, S.: Assessment of energy system management—the case study of Kazakhstan. E3S Web of Conferences, vol. 135, p. 04056 (2019). https://doi.org/10.1051/e3sconf/201913504056
21. Demessinova, A., Saparbayev, A., Seidakhmetov, M., Kydyrova, Z., Onlasynov, E., Shadiyeva, A., Daurbayeva, M.: Renewable energy sector of Kazakhstan: factors of its sustainable development. Int. J. Manag. Bus. Res. **8**(1), 34–51 (2018)

Networks, Infrastructure and Supporting Services in Innovations Development

Development of Synchromodal Logistics Based on Modern Technologies

Bauyrzhan Yedilbayev, Tatyana Sizova, Aikorkem Nurlankyzy, and Almagul Kazmagambetova

Abstract As supply chain management is becoming demand driven, logistics service providers need to use real-time information efficiently and integrate new technologies into their business. Synchromodal logistics has emerged recently to improve flexibility in supply chains, cooperation among stakeholders, and utilization of resources. We survey the existing scientific literature and real-life developments on synchromodality. We focus on the critical success factors of synchromodality and six categories of enabling technologies. We identify open research issues and propose the introduction of a new stakeholder, which takes on the role of orchestrator.

Keywords Synchromodality · Logistics service providers · ICT/ITS technologies · Fifth-party logistics · Integration platform · Implementation strategy of digital market · Advertising

1 Introduction

As supply chain management becomes more demand-driven, logistics service providers must effectively use real-time information and integrate new technologies into their business. Synchromodal logistics has recently been introduced to improve

B. Yedilbayev (✉)
Al-Farabi Kazakh National University, al-Farabi Ave. 71, 050040 Almaty, Republic of Kazakhstan
e-mail: yedilbayev.b@kaznu.kz

T. Sizova
Plekhanov Russian University of Economics, 117997, 36, Stremyanniy Moscow, Russia

A. Nurlankyzy
Atyrau PetroTehnic High College, Baimukhanov st.45, Atyrau, Republic of Kazakhstan

A. Kazmagambetova
Asfendiyarov Kazakh National Medical University, Tole bi st.94, 050000 Almaty, Republic of Kazakhstan

G. Mutanov and A. Serikbekuly (eds.), *Digital Transformation in Sustainable Value Chains and Innovative Infrastructures*, Studies in Systems, Decision and Control 443,
https://doi.org/10.1007/978-3-031-07067-9_8

flexibility in supply chains, collaboration between stakeholders, and resource utilization. We study the existing scientific literature and real developments on synchromodality. Moreover, the shape of the product itself changes, i.e., the product itself changes. The physical nature of goods, services, and information produced by the company to create consumer value. Information and communication technologies permeate the entire value chain at each link, changing the way activities are performed and the links between them. Customers' needs are more fully met, products are given a new look or shape, and the scale of competition changes. Based on these key points, it becomes clear why information technologies are of such strategic importance and how they differ from other technologies used in business. With the development and spread of information technologies and systems, consumers have gained access to databases of available goods and services, and the ability and right to choose the best of them. Internet technologies create new sources of value and shareholder value [1].

Value chains are disintegrating and industries are being reformed. The number of specialized firms is growing. There are positive changes in product distribution channels. New intermediaries are created and old ones are eliminated. Additional value for consumers is provided by the fact that working through the Internet allows customers to receive goods and services adapted to their individual requirements. It becomes possible for consumers to order technical devices, systems, and software products. A special feature of the digital revolution is that not only content but also processes are being translated into digital form. In particular, digital marketing processes are being actively implemented. For example, this is marketing research using Internet technologies and big data Analytics; managing customer relationships through. One of the biggest changes in human interaction is the recent proliferation of social networks. The rapid growth of web platforms that facilitate social behavior on the Internet has significantly changed the nature of human activity, habitat, and interaction [2].

Real social relationships were transferred to the virtual world, which led to the creation of online communities that bring together people from all over the world. This movement into the digital dimension allows people to share knowledge, entertain each other, and foster dialogue between different cultures [3].

The peculiarity of such advertising is that half of the cost of a click is taken by the advertising network, and the second part of the cost of a click is received by the webmaster-the Creator of the site where the ads are broadcast. These ads can be similar in subject matter (close to the page context), or different from it. This feature of setting up an ad campaign is called "behavioral targeting". If it is disabled, the site will only display ads that match its content. We focus on critical success factors of synchromodality and six categories of enabling technologies. We identify open research problems and suggest introducing a new stakeholder who takes on the role of an organizer to coordinate and deliver services through the technology platform [4].

2 Literature Review

Synchromodality is an evolving and attractive concept in logistics that has been developed and implemented in the Benelux region over the past decade. The main goal of synchromodality is to reduce costs, emissions, and delivery times while maintaining the quality of supply chain service through rational use of available resources and synchronization of transport flows. The implementation of the synchromodality concept and some research projects based on this practice have already shown how various types of logistics tasks can be achieved or significantly improved, including avoiding empty capacity, responding to failures, and reducing freight transport in favor of Railways, ships, and barges [5].

However, modern literature lacks a clear, complete, and generally accepted definition of synchromodal logistics and its features. By our definition, synchromodality is the provision of efficient, reliable, flexible and sustainable services through coordination and collaboration of stakeholders and synchronization of operations within one or more supply chains managed by information and communication technologies (ICT) and intelligent transport system (its) technologies [6].

"Logistics companies should become partners for the client, stay ahead of trends and offer innovative solutions. Simply lowering rates by reducing margins is not possible, because in addition to increasing customer requirements, the environment in which transport companies operate is also changing," says Alexandra Kotsemba, intermodal transport Manager at Raben Transport.

The characteristics discussed in our definition have not yet been described together, and a common vision for how to use them to support new trends in supply chain management does not yet exist. The purpose of this study is to provide a detailed understanding of this definition, based on a critical review of the existing scientific literature and actual developments in the field of synchromodal logistics. In fact, our analysis goes beyond a simple overview of the state of this transport system.

Agnieszka Pawlowska-Ogrodnik from the waterproofing logistics and customer service Department at Renolit claims that we have been using our offer in the field of intermodal transport for a year now. In terms of traditional offers, this is a very attractive offer, because synchro modular solutions allow us to get real savings. Especially with a large volume of cargo, which we are talking about in our case [7].

According to Ziyadin, after analyzing the GEO for synchromodality at various levels of decision-making, we identify the relevant stakeholders and auxiliary technologies necessary for the implementation of synchromodality in practice. Finally, we provide recommendations for reviewing CSR for each group of enabling technologies (outlining open research issues) and possible synergies achievable on a common platform. Since stakeholders need coordination, we also identify an organizer (fifth-party logistics, 5PL) who can get this role to provide and manage the shared platform [8].

Maddison, A. writes that despite its importance, synchromodality is at an early stage both from a scientific and practical point of view. The existing materials are few and relate only to one or more aspects of the issue. A lot of material can be found on the Internet, but it is rather disordered. Instead, in our study, we look at the phenomenon of synchromodal logistics through a comprehensive and systematic analysis, ranging from social and business consequences to operational management and enabling technologies. Thanks to the features of our analysis, our work provides insight for both managers involved in logistics operations and management, as well as for companies developing technological solutions or services. In addition, the study is a useful source of information for researchers and practitioners in various fields who are interested in this new logistics paradigm [9].

3 Methodology

For completeness, we will explain the method used in this study. Our goal is to collect, study, and classify existing literature related to synchromodality, clarify the current state, and create a solid Foundation for future research on this topic. We do not compare specific analytical models but discuss existing approaches at different levels. In fact, we analyze the synchromodal logistics paradigm from a broad perspective, while analytical models usually handle specific aspects (for example, operational optimization tasks). The main sources for this work were ransacked in the major academic databases and web resources (e.g., Scopus, Google scholar and research gate) using the selected logistics key words (such as synchromodality, synchromodal, synchro-mechanism, synchro-modal, intermodality, multimodality, 5pl LLC, the Fifth party logistics provider, 4PL, fourth party logistics provider, 3PL, third-party organization, organization of data analysis, data analysis, supply chain, Industry 4.0). In addition, we include online reports on case studies or research projects in order to present real-world synchromodality applications and advances in enabling technologies [10].

3.1 Communication Organization

The supply chain is "the flow of processes for moving goods from the customer's order through the raw material, delivery, production, and distribution stage of products to the customer", and supply chain management is "managing the chain of events in this process". Synchromodality is often associated only with the synchronization of transport modes and thus covers only the relevant supply chain management activities. However, synchromodality has the potential to address multiple supply chain activities in addition to transport. Therefore, in our opinion, synchromodal logistics should be considered in a broader sense, perhaps assuming coordination of all relevant operations in supply chain management [11].

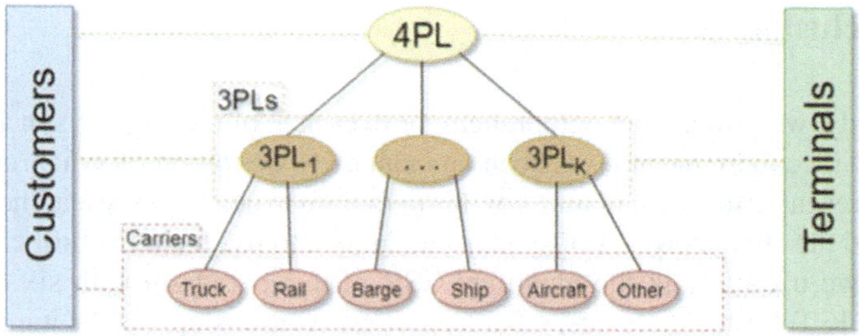

Fig. 1 4PL Logistics services

The context in which synchromodality works is presented in the following subsections. In particular, we represent stakeholders involved in synchromodal logistics, highlighting their current interaction . In addition, we show that such interactions are not well suited to effectively respond to new trends in supply chain management.

Stakeholders and their relationships [12]. The evolution of supply chain management has been gradual, moving from relationships based on simple dyadic relationships to complex multi-factor networks. Currently, participants in the supply chain are linked directly or indirectly, and their relationships are not only related to the physical movement of goods, but also to the exchange of information. When stakeholders build supply chains based on such complex networks, there is a need for a coordinating intermediary to achieve a balance between cooperation and competition. In addition, stakeholders are increasingly aware that their own success is closely linked to the effectiveness of the entire supply chain. For these reasons, modern businesses rely heavily on so-called logistics service providers (LSP), which offer their customers a wide range of services, from simple transport operations to complete supply chain management. In fact, depending on their needs and convenience, companies decide to outsource some activities or the entire logistics of their LSP supply chain [13] (Fig. 1).

It is important to note that the most common types of outsourcing are 3PL services, such as transport, warehousing, and freight forwarding. In contrast, 4PL services are relatively new in the real market, and their advantages are still not recognized by many companies that are often reluctant to give full control of their supply chain to outsiders [14]. However, since Internet service providers often offer different types of services, it is not always possible to accurately classify real companies into one of the previously analyzed categories [15].

The previous description reflects the current situation, but new trends in supply chain management create a need to rethink and redesign the organization and stakeholder engagement [16].

4 Conclusion

In this study, we presented a comprehensive overview of the state of synchromodal logistics. Since there has not yet been an exhaustive definition of synchromodality, we have presented the main features of this paradigm to help other researchers clearly understand the characteristics of synchromodality and use them for further study. In addition, we tried to link the identified CSR of this paradigm with six classes of providing ICT / it technologies. Characteristics, relationships, and future research directions were analyzed for each technology that provides support. In addition, we noted the importance of enhancing synergies between enabling technologies by defining a common technology platform that allows stakeholders to synchronize their activities and simplify their business. We also proposed a new type of LSP (5PL, often discussed in the literature but never clearly defined) that could develop this common platform by taking on the role of a synchromodality Orchestrator [17].

Our definition of synchromodality, given in the introduction, is based on the main characteristics identified in the analysis of this study. Now we motivate our choice. With "efficient, reliable, flexible and sustainable services", we focus on the characteristics that synchromodal services should have, inheriting efficiency and sustainability from previous concepts and implementing reliability and flexibility to provide fault-tolerant services. In the "stakeholder coordination and collaboration" section, we emphasize the importance of collaboration and the need for an organizer (5PL) to coordinate the system. The main feature of synchromodality is laid down in the "synchronization of operations". Note that when using the word operations, we move the focus from syncing only transport operations to possibly syncing all operations that are performed in the supply chain. When we refer to "one or more supply chains", we point out that collaboration can be improved by integrating activities and related information into a broader network. Finally, "driven by ICT/IT technologies" emphasizes the importance of the described technological mechanism for the effective implementation of synchromodal logistics concepts [18].

In conclusion, we argue that synchromodal logistics should be explored in the future by considering all possible assistive technologies related to CSFs. In particular, scientists analyzing each of these technologies should develop studies that confirm the features of the synchromodal paradigm. In addition, integration between technologies within the 5PL common platform is important to increase the benefits of using synchromodality and building a stronger, more reliable, and flexible system. In addition, it is important to examine and define the role of 5PL service providers in the supply chain from various perspectives, such as business management or law. 5PL service providers can be interpreted either as technology solution providers, or as an evolution of 4PL service providers with their businesses largely driven by technology. There are good reasons to use recent or current research projects and case studies of real-world applications of synchromodal logistics as a starting point for implementing synchromodality in a broader context. For example, completed a feasibility study for implementing synchromodality in a developing country (Ghana). However, building a global supply chain based on synchromodality requires additional effort.

In addition, social media offers the potential to create awareness and interest through viral or rapid dissemination of experiences and opinions about products and services. Digital marketing is based on Internet channels with the highest frequency of use, which are dynamic, they can change from year to year and are always influenced by market trends. However, this type of advertising does not work in all business niches and requires special knowledge to get a high return on it [19–21].

If you are new to this topic, a directologist who specializes in contextual advertising will help you set up your ad campaign professionally.

This ad can be used as the only tool to attract customers on the Internet, or by combining it with other types [22, 23].

References

1. Brouwer, E., Kleinknecht, A.: Determinants of innovation: a microeconometric analysis of three different innovation output indicators. In: Kleinknecht, A. (ed.) Determinants of Innovation: The Message from New Indicators, pp. 99–124. Macmillan, London (2016)
2. Dosi, G.: Sources, procedures and microeconomic effects of innovation. J. Econ. Lit. **26**, 1120–1171 (2008)
3. European Commission European Economy. 1996 Broad Economic Guidelines, No. 62, pp. 128–13 1. Brussels/Luxembourg (2000)
4. Feldman, M.: The Geography of Innovation. Kluwer, Dordrecht (2014)
5. Jaffe, A.B.: Technological opportunity and spillovers of R&D: evidence from firms' patents, profits and market value. Am. Econ. Rev. **76**, 984–1001 (2006)
6. Krugman, P.: Pop Internationalism. MIT Press, Cambridge (MA) (2016)
7. Lindblad, J.T., van Zanden, J.L.: Dutch foreign trade, 1872–1913' (in Dutch). Economischen Sociaal Jaarboek **52**, 231–269
8. Ziyadin, S., Serikbek, A.: Internet-Marketing in the sphere of higher education. In: International Scientific Conference "Digital Transformation of the Economy: Challenges, Trends, New Opportunities", pp. 290–297. Springer, Cham (2019)
9. Maddison, A.: Dynamic Forces in Capitalist Development. Oxford, OUP (2001)
10. Mensink, N., van Bergeijk, P.A.G.: 'Globlahblah' (in Dutch), Economisch Statistische Berichten **81**, 914–916 (1996)
11. Nooteboom, B.: Towards a dynamic theory of transactions. J. Evol. Econ. **2**, 281–299 (2000)
12. Nooteboom, B.: 'Innoveren, globaliseren' (in Dutch). Economisch Statistische Berichten **81**, 828–830 (2002)
13. Patel, P., Pavitt, K.: Large firms in the production of the world's technology: an important case of non-globalization. J. Int. Bus. Stud. **22**, pp. l–21 (2000)
14. Ruigrok, W., Van Tulder, T., The Logic of International Restructuring. Routledge, London (2013)
15. Rosan, C.D.: Can PlaNYC make New York City "greener and greater" for everyone?: sustainability planning and the promise of environmental justice. Local Environment **17**(9), 959–976 (2012)
16. Wang K., Yu. S., Zhang V.: Regional energy and environmental efficiency of China: dynamic assessment based on DEA analysis. Math. Comput. Model. **58**(5–6), 1117–1127 (2013)
17. Fahimnia, B. et al.: The role of green logistics and transport in sustainable supply chains. Green Logistics and Transport, 1–12. Springer, Cham (2015)
18. Teter, J. et al.: The Future of Trucks: Implications for Energy and The Environment (2017)
19. Royle, J., Laing, A.: The digital marketing skills gap: developing a Digital Marketer Model for the communication industries. Int. J. Inf. Manage. **34**(2), 65–73 (2014)

20. Negoiță, O.D., Purcărea, A.A., Popescu, M.A.M.: Research on online promoting methods used in a technological society. Proc. Manuf. **32**, 1043–1050 (2019)
21. Stephen, A.T.: The role of digital and social media marketing in consumer behavior. Curr. Opin. Psychol. **10**, 17–21 (2016)
22. Wertime, K., Fenwick, I.: Digital Marketing: The Essential Guide to New Media and Digital Marketing. John Wiley & Sons (2011)
23. Mullakhmetov, K.S., Sadriev, R.D., Krotkova, E.V., Gabaidullina, L.A., Akhmetshin, E.M.: Control transformation in the lean management system. In: Paper presented at the Proceedings of the 32nd International Business Information Management Association Conference, pp. 1717–1723. IBIMA 2018—Vision 2020: Sustainable Economic Development and Application of Innovation Management from Regional Expansion to Global Growth (2019)

Staff Threats and HR-Strategy: How Talent-Management can Provide the Surviving of Companies During the Industry 4.0

Elena Lysenko, Elena Makhalova, and Irina Plotnikova

Abstract This article is dedicated to talent-management as a modern instrument for ensuring success of organizations within the Industry 4.0. The cardinal attention is given to the threats of the 4th Industrial Revolution to HR-sphere in the future and to the role of talent-management that can prevent them. The other focus is on demonstrating the actual statistics that point to direct correlation between talent-management and efficiency of an organization in the future. In the present desk research modeling, analysis and synthesis, generalization, document analysis and induction were used. Analysis of two main directions such as presence of personnel threats of Industry 4.0. (1st) and HR-strategies which help to overcome personnel threats (2nd) showed that they will entail large financial and energy costs due to the training and incentives provided for talented employees. However, with the onset of the Fourth Industrial Revolution, organizations will gain much more than lose. Talent-management will become a significant factor for company's success. Therefore, a right approach to talent-management is quite justified and serious investment. It helps the company to recruit and retain qualified and motivated employees who are able to realize long-term objectives of the organization.

Keywords Industry 4.0 · Human resources · Talent management · HR-strategy · Staff threats

1 Introduction

Talent Management has become in much demand for organizations in fast changing modern world. Many executives realize that "talented" employees are those who can increase the competitiveness in the market. However, "talented" employees require much more attention and it can be not easy to retain them in the company. Numerous research papers devoted to the use of talent-management in the framework of the

E. Lysenko (✉) · E. Makhalova · I. Plotnikova
Ural Federal University named after the First President of Russia B. N. Yeltsin, Mira str. 19, 620002 Yekaterinburg, Russian Federation
e-mail: e.v.lysenko@urfu.ru

G. Mutanov and A. Serikbekuly (eds.), *Digital Transformation in Sustainable Value Chains and Innovative Infrastructures*, Studies in Systems, Decision and Control 443,
https://doi.org/10.1007/978-3-031-07067-9_9

Industry 4.0 has been published in foreign editions, though in Russian science the present topic is not developed enough.

There is little doubt that the attraction, development, and retention of talents are nowadays one of the most critical challenges faced by companies worldwide. Despite the increasing scholarly attention during the last years many questions remain, particularly, those related to how (and why) talent management is conceived, implemented and developed within organizations, not to mention about its outcomes or effectiveness [1].

Taking into consideration The Fourth Industrial Revolution, HR-managers should understand the necessity to use adequate and suitable HR management strategies. HR-management should be more strict but flexible in order to reflect the market dynamics and inform employees of changes. With the development of the Industry 4.0, talent-management is becoming essential for successful activity of an organization.

Organizational talent management has been widely recognized as a key driver of firm performance [2]. Existing theoretical and empirical work in the domain has drawn on Social Exchange Theory to suggest that talent management affects organizational performance by eliciting positive reactions such as high organizational commitment and work effort among the firm's most high performing and high potential employees (i.e., talented employees).

2 Literature Review

Talent-management uniqueness. The term "talent-management" is to be considered from different angles. For instance, Egorova understands talent-management as "the term in modern theory and practice of management denoting a new concept in the interaction with employees characterized by high potential enabling them to reach exceptional results in labor activity [3]. Ed Michaels, H. Handfield-Jones and E. Axelrod consider talent-management as "the process to attract attention, conduct selection; stimulate and develop creative employees representing the principal company's asset. It is their creativity that helps the company to rocket". This management type is based on the idea that "the market leaders are the companies which actively look for and attract the most talented employees and then develop their creative potential and contribute to their self-realization" [4]. Odegov writes that "talent-management can be considered a part of HR-management, aimed at attraction of highly qualified employees, integration of new employees as well as retention of personnel exploiting their valuable skills and professional knowledge for current and future business projects" [5]. To summer up, it should be noted that talent-management is an absolutely new approach in HR-management aimed at creation of the unique system of HR-management. Thanks to HR-management a corporation is able to increase its effectiveness and can fully participate in competition. Essentially, firms now have to compete intensely for the scarce human capital [6]. The task of a modern organization is to cultivate a culture of talent using various methods [7]: 'Resource Based View (RBV)', 'Managing expert talent', 'Supply chain approach to talent management", 'Employer branding and career management', 'Typology of management strategies' etc.

Industry 4.0 as a digital technical revolution. The launch of a new digital industrial technology known as the Industry 4.0, representing a transformation, allows collecting and analyzing the data at various machines. This provides quicker, more flexible and more effective processes for the production of the goods of better quality at lower prices [8]. The Industry 4.0—is a term first used in 2011 at the Hannover fair by the group representing the German Industry in the framework of the initiative on the increase of competitiveness of Germany in the fast internationalization and cybernetization of production. The Fourth Industrial Revolution challenges the traditional way of production system functioning at the factories with the centralized and autonomous systems not connected among them. Revolutionary changes will soon lead to "smart factories" with the option to self-manage problems and internal processes. In the framework of production system, the technologies are becoming more complex and acquire the title of "cyber-physical production systems". Such devices are able to diagnose themselves, "predict" faults and inform the user about the need for maintenance. Many devices already have some options for self-diagnosis. For instance, a phone can inform the owner about the need to change the battery if it is worn; a smart printer will notify the owner about the low ink level in cartridge and the necessity to change it soon by sending a notification to the application in the phone [9]. Besides it, as Whysall et al. [10] emphasize, the transformational changes to business environments brought about by the fourth industrial revolution create a perfect storm for strategic human resource management, prompting a need to explore the implications of this context for talent management theory and practice. The speed of technological change brought about by Industry 4.0 had created a significant gap between current capability of employees and the rapidly evolving requirements of their roles, prompting a need to consider new and more effective approaches to talent development.

3 Methodology

Methodology of research is based on analyses of open sources of analytical institutes and consult agencies which consider global trends of Industry 4.0 in the aspects of transformations in the area of Human resources management especially in talent-management as a modern HR-trend. The Desk study was conducted with using of modeling, analysis and synthesis, generalization, document analysis and induction, were used. Our task was to evaluate 2 main directions:

1st—presence of personnel threats of Industry 4.0.;
2nd—HR-strategies which help to overcome personnel threats.

Expected result is to establish the role of talent-management in the as a mode to provide the surviving of companies during the industry 4.0.

4 Results

Personnel threats of Industry 4.0. Every past technological revolution (The First, The Second and The Third Industrial Revolutions), as a result, created more jobs than it destroyed, thus they provided some significant benefits: from the increase of level and length of life to production and economical growth. Nevertheless, many people are scared that this time these changes can be so rapid and huge, and their influence so uneven and destructive that not only individual means of existence but the stability of the society itself could be put under risk.

The following threats of Industry 4.0 can be singled out.

Threat 1: fears that human labor will be replaced by information innovations. The employer may introduce technologies able to perform the role of employees more effectively but at a lower cost.

Experts [11, 12] indicate that the reduction in the number of jobs in the future may lead to some serious threats to the economic security of the state, such as the increasing polarization of incomes, which may result in increasing social tensions, the disappearance of a number of professions, which will inevitably lead to a rise in the unemployment rate, which, in turn, may contribute to an increase in crime.

One should realize that the fears are not groundless. Less qualified jobs requiring simple and repetitive actions will be mostly substituted by the intellectual cyber-systems. At the same time, the employer experiences more difficulties to find talented employees with the required competences. That is, at the present time there is much inconsistency between the need of the employer and the skills and knowledge the talented applicants possess.

However, a number of foreign employees provide the results of their research in the sphere of changes in the market in the framework of the Industry 4.0. This well neglects the present threat. In particular:

1. The head economist PwC John Hawksworth [13] and his team analyzed 15 million working places in 20 countries. They came to a conclusion that the *increase of automatization of working process creates more working places than there are now* (Fig. 1).
2. International recruiting company Hays in its report «The Hays Global Skills Index 2019/20» has presented the result of the analysis of the markets in 33 countries. In their opinion, *improving the technologies will lead to the loss of jobs and, as a result, many low-qualified jobs will disappear. The innovations will also reduce the production cost of products and services, what in its turn will increase the consumer's purchase power. These additional costs can be a financial base for the creation of new jobs* [14].
3. McKinsey Global Institute expects 200 jobs to appear by 2030, so *this growth may successfully compensate the losses of jobs occurred as a result of automatization* [15].
4. The representatives of The Economist in their article supposed that 50% of jobs are vulnerable to automation. However, some industries are more likely to be

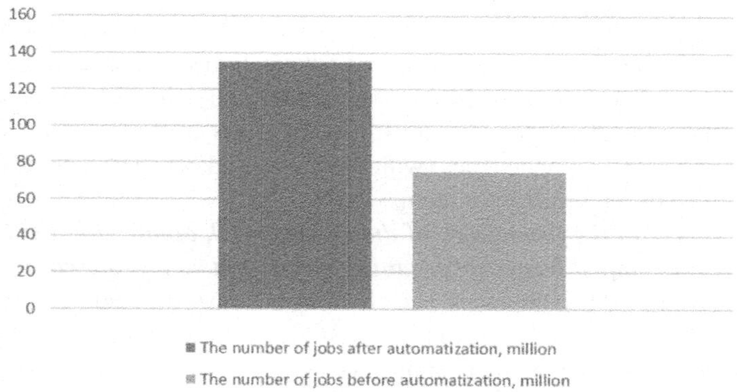

Fig. 1 The impact of automatization on jobs (*Source* Hawksworth [13])

modernized than the others. *Employees whose jobs require communicative skills and knowledge will not be automatized in the nearest future* [16].

5. The Atlas of New Professions by The Agency for Strategic Initiatives (ASI) developed in collaboration with Skolkovo [17], BBC News [18] and World Economic Forum [19] provide statistics according to which in the future many types of activities that are numerous and popular in today's Russia, such as: accountant, banking operator, public services operator, etc., will cease to exist. However, new professions are likely to appear, such as: medical robot operator, city farmer, production equipment supervisor, etc. The editions provide some statistics of some new professions and the impact of automatization on them (Fig. 2).

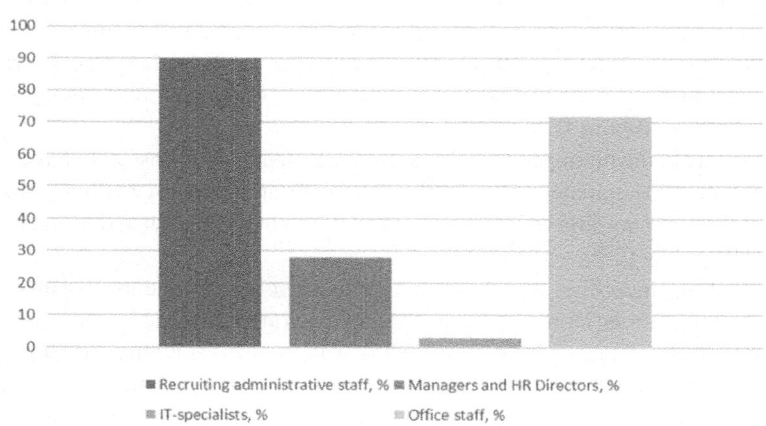

Fig. 2 Probability of disappearance of professions (*Source* Prepared by the authors and is based on the following sources: Luksha et al. [17]; Stylianou [18]; Martin and Savchenko [19])

Thereby, it may be concluded that industry 4.0 will be developing according to the scenario of moderate automatization and robotization with preservation of domination of human labor in most business processes and spheres of economy [20]. Communications with people will constitute the basis of the activities of digital personnel, and social competencies (with obvious significance of technical competencies) will be of top priority for them.

Threat 2: within the framework of the Industry 4.0, along with digitization of production, the competitiveness of organizations automatically increases, which, in some ways, is a threat to companies. Headhunting becomes the leading way to search for personnel that is why retention of talented staff may be quite a challenging task for HR-managers. To succeed in this, employers should realize what exactly talents the company needs not only in short term run but in a long term too [21]. The employer should possess a clear management strategy corresponding with the business strategy of the whole organization.

Companies´ competitiveness is crucial for their sustainability in the medium and long term. Much of their performance depends on their human resources, valuable strategic asset with relevant role in transforming inputs into fruitful outputs. Nowadays innovation depends less on physical assets and more on intangible assets. In a world of change, companies need human resources able to adapt to environment and its competitiveness depends on its ability to have talented workers. The biggest challenge of human resource managers is precisely be able to attract, retain and develop talent. Companies that do a better talents management are those that have better results in terms of innovation. The power those sustainable companies have in attracting and retaining their talent will depend on the mechanisms they are able to create to maintain and retain them, proactively acting on their attraction and avoiding voluntary departure. Companies must act on the dimensions of employer branding, recruitment and selection. They must be attentive to the economic and social environment and develop policies of retention, namely as linked with compensation and benefits, leadership style, performance management, carrier management. Training and development, climate and organizational culture, organizational communication and working conditions [22].

According to the findings of the research conducted by McKinsey Global Institute and based on 500 companies, 96% of the executives are not ready to digital changes. Many of them are not able to survive the competition in the market and do not possess a clear HR policy [23].

Curzi et al. underline, that in digital competitive environments, organizations' ability to innovate is more than ever the key to competitive advantage. One way to cope with this increased pressure for innovation is to capitalize on employees' ability to generate new ideas and use these as building blocks for new and better products, services, and work processes. Individual innovation thus emerges as a key competence required from workers, in turn crucially affecting the way managers make employees contribute to organizational goals and assess their performance [24].

As shown by Bogoviz et al. [25] in addition to the competence "to generate innovation" in the development of "knowledge economy" and Industry 4.0 professional competence "computer programming" and "digital thinking" will be in demand, which will require moving away from a narrow approach to training to establish a broad specialization of workers.

In order to prevent the above mentioned threats, companies should undertake some actions using talent-management in order to survive in the Industry 4,0., which include the following:

1. *Continuous training of talented staff.* The training may be carried out in two different ways. The first way—a detailed review of the innovations of digital transformation of the organization. The second way—explanation why these changes are essential. With the lack of qualitative continuous training, personal development of talented employees within the framework of the Industry 4.0 may appear quite troublesome. The lack of applicants with essential skills and competences for a particular sphere in the job market makes employers start providing training programs for talented personnel already working in the company, or, if necessary, for personnel with high potential and well developed related skills. Although digitalization proves to be a problem for a company, as technology and methodology are continuously changing, this step may contribute to solving one of the threats stated above.

   World Economic Forum and The Manufacturer in their researches provide statistics according to which 35% of skills of employees will have dramatically changed by 2030 [26] (Fig. 3).

   In addition, 71% of executives believe that professional training is becoming a real alternative to higher education. At the same time, a trainee must learn from highly qualified and experienced employees [27] (Fig. 4).
2. *Retention and stimulating of talented employees.* Universal ways not suitable for talented employees, but proved to be highly profitable for the company, can be used for retention and stimulation of mid-level staff. Talented employees require a special attitude, and the following tools can be used to stimulate and retain them, in particular:

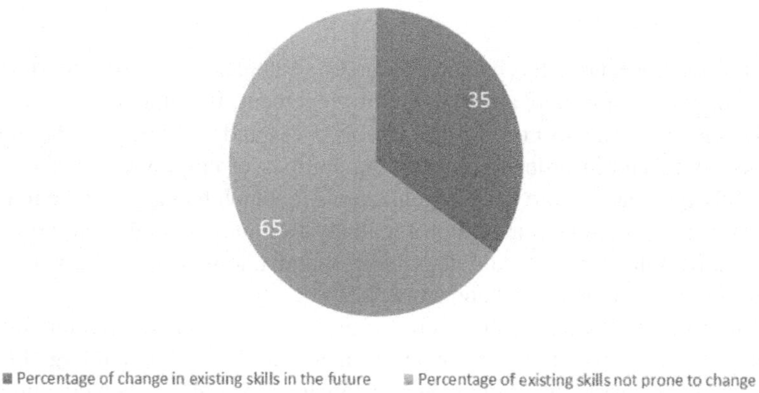

Fig. 3 Probability of changes in employee skills (*Source* Schwab and Samans. The Future of Jobs [26])

Fig. 4 Managers' opinion on the role of higher education in professional activity (*Source* Coulthard. Annual Manufacturing Report 2019: Taking the pulse of British industry [27])

- *building an effective career ladder*: regular review of the organizational structure, promotion of talented employees;
- *paying attention to any idea expressed by a talented employee*: quite often the ideas offered by talented employees can be described non-standard and rather creative, and if carefully considered, many of them prove to be profitable;
- *opportunity for international internship:* every year several talented employees should have a possibility to gain new experience in foreign companies;
- *"incentive tourism"*: spending time outside the organization for the purpose of corporate leisure, relaxation, training and team building;
- *career coaching*: work with inner motivation, identifying the potential of a talented employee and determining the ways of its implementation. As a result, the feeling of self-satisfaction of a talented employee increases; a particular environment contributing to their retention and attraction is formed in the organization [28].

Therefore, nowadays, the human resources strategies are considered to be the main components of organizations to get improvement. It is imperative to consider the strategies, in order to increase the effectiveness and efficiency of the management activities, and in order to develop the abilities of employees. In this regard, successful organizations are those organizations in which managers and employees, are always in a dynamic competition for innovation and creativity on the basis of their organization strategies, and also those organizations in with thinking among the forces have become a habit and a task [29].

HR-strategies. Having carefully studied the methods of preventing threats to the Industry 4.0 mentioned above, it can be assumed that they will entail large financial and energy costs due to the training and incentives provided for talented employees. However, with the onset of the Fourth Industrial Revolution, organizations will gain much more than lose.

This statement is proved by a number of authoritative studies. According to Accenture, one of the leaders on the market of professional services and digital technologies, an average American company can save 85,000 US dollars per an employee in case it implements five technologies:

1. AR/VR;
2. autonomous vehicles;
3. database;
4. machine learning;
5. the development of mobile communications [30].

Baykal notes, that digitalization influenced on HRM-strategy: "Owing to the digital revolution, e-HR emerged as a new way of practicing HRM activities with the latest web-based and computer-based tools and applications. These applications eased the work of HR professionals and served them the opportunity to focus on their core work, namely strategic human resources activities rather than procedural paperwork of the department" [31].

The studies carried out by McKinsey Global Institute show that 15 trillion dollars of salary worldwide can be saved by 2030 by intellectual automatization technologies [32].

In 2014 and 2015 such research associations as IFIDZ & FAZ-Institute and Engeser provided the results of their studies on the changes of work caused by digitalization. The following factors were called key elements of work in the future: teamwork (85% of respondents'), customer ideas in product development (80%), partnerships (78%), and flexibility of work schedule (80%). Finally, 77% of all managers surveyed confirmed that they would feel the pressure of digitalization (Fig. 5) [33].

Brandi et al. [34] adds, that as organizational boundaries are blurred because of the complexity due to globalization, digitalization, and demographic changes, in practice, new ways of organizing work can be found that depend on the specifics of

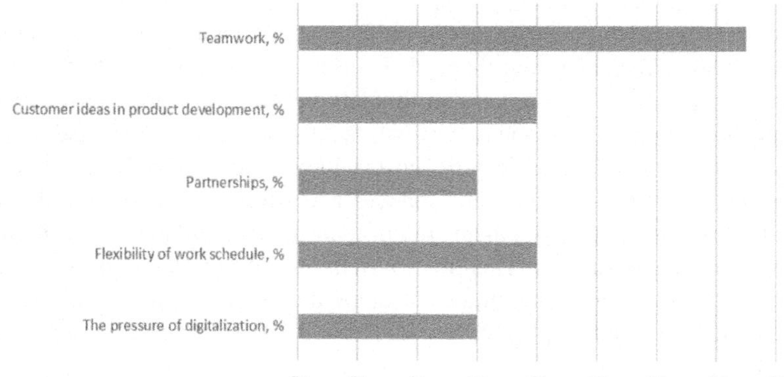

Fig. 5 The key elements in the work of the future (*Source* Bonekamp, Consequences of Industry 4.0 on Human Labour and Work Organisation [33])

the work situation. So, different modes of coordination in employment are applied in a fluctuating manner that depends on the specific situations.

The Fraunhofer IAO Research Institute in Germany prepared a survey of 518 representatives of industrial organizations about their views on the effects of Industry 4.0 on manufacturing. 51% of respondents said that they expect manual labor to reduce, at the same time 54% of respondents expect more control and assessment. 86% of the participants emphasized the increasing importance of continuous training. At the same time, 77% expect more attention to be paid to interdisciplinary cooperation, but 67% think that the standards of IT-competency will increase. The scientists from Fraunhofer believe that jobs are going to decrease as the result of cyberphysical systems, but they do don suppose that the scenario according to which factories will be able to operate without any human interference [33].

5 Conclusion

The effects of the Industry 4.0 may be quite destructive for companies which are not ready for changes. Critical to a corporation's growth and prosperity is gaining and retaining competitive advantage. Although corporations may pursue many paths to this end, one that is frequently not recognized is capitalizing on superior human resource management. Currently, many companies recognize the growing importance of their human resources, but few are conceptualizing them in strategic terms—in ways to gain a competitive advantage. As a result, many companies forego the opportunity to seize competitive advantage through human resource practice initiatives [35].

It is profitable for managers to use the instruments of talent-management to occupy and hold a stable position on the labor market. In the context of strategic renewal, talent management process consists of identification of key projects to address critical business opportunities and challenges, the identification of talented employees to execute them, and the identification and creation of key positions [36].

Those managers who spent much on continuous training, incentives and retention of talented employees may find themselves in the situation of a budget surplus in the future. It should be also noted that this HR-strategy has a long-term company development perspective. At the same time, the executives not interested in talent-management because of "unnecessary" costs will be successful only in a short-term run because of their unreadiness for changes and strong belief that the threats of the Industry 4.0 do not concern them. Talent-management will become a significant factor for company's success. Therefore, a right approach to talent-management is quite justified and serious investment. It helps the company to recruit and retain qualified and motivated employees who are able to realize long-term objectives of the organization.

Acknowledgements The work was supported by Russian Foundation for Basic Research (RFBR) under Grant № 19-010-00705 'Development of tools for assessing the impact of social pollution of labor relations on the employees' well-being in a digital economy'.

References

1. Gallardo-Gallardo, E., Thunnissen, M., Scullion, H.: Talent management: context matters. Int. J. Hum. Resour. Manag. **31**(4), 457–473 (2020). https://doi.org/10.1080/09585192.2019.1642645
2. Meyers, M.C.: The neglected role of talent proactivity: integrating proactive behavior into talent-management theorizing. Hum. Resour. Manag. Rev. **30**(2) (2020). https://doi.org/10.1016/j.hrmr.2019.100703
3. Egorova, I.A.: Kontseptsia upravlenia talantami: osnovnie podhodi [The concept of talent management: main approaches]. http://cyberleninka.ru/article/n/ kontseptsiya-upravleniya-talantami-osnovnye-podhody. Last accessed 19 Nov 2019
4. Voina za talanty [Tekst]: perevod s angliiskogo / E. Maikls, H. Henfild-Djons, E. Ekselrod. - Moskva: Mann, Ivanov i Ferber, (2011)
5. Odegov, Y.G.: Upravlenie personalom: uchebnik dlya bakalavrov [HR-management: guidebook for bachelors]. Uright, Moscow (2016)
6. Amankwah-Amoaha, J., Debrahb, Y.A.: Competing for scarce talent in a liberalised environment: evidence from the aviation industry in Africa. Int. J. Hum. Resour. Manag. **22**(17), 3565–3581 (2011). https://doi.org/10.1080/09585192.2011.606111
7. Shet, S.V.: Strategic talent management–contemporary issues in international context. Hum. Resour. Dev. Int. **23**(1), 98–102. https://doi.org/10.1080/13678868.2019.1650324
8. Skalabre, O.: Vovlechenie Industrii 4.0 i protses rosta [Implementation of Industry 4.0 and growth]. http://absel.ua/news/vovlechenie-industrii-4-0-i-process-rosta.html. Last accessed 14 Oct 2019
9. Kozlov, D.R.: Industria 4.0 Perspektivi i vizovi [Industry 4.0: perspectives and challenges]. http://cyberleninka.ru/article/n/industriya-4-0-perspektivy-i-vyzovy. Last accessed 15 Dec 2019
10. Whysall, Z., Owtram, M., Brittain, S.: The new talent management challenges of industry 4.0. J. Manag. Dev. **38**(2), 118–129 (2019). https://doi.org/10.1108/JMD-06-2018-0181
11. Kuchumov, A.V., Pecheritsa, E.V., Kirillov, A.T., Chaikovskaya, A.V.: Current problems of the higher education development in the Russian Federation considering the digitalization processes. In: Paper presented at the ACM International Conference Proceeding Series (2019). https://doi.org/10.1145/3372177.3373341
12. Fedorova, A.: Dealing with toxic leadership: empirically defining and evaluating destructive labour relations. In: Mesquita, A., Mesquita; Silva, P. (eds.) Proceedings of the 15th European Conference on Management, Leadership and Governance, ECMLG 2019. Academic Conferences and Publishing International Limited, pp. 133–140 (2019)
13. Hawksworth, J.: AI and robots could create as many jobs as they displace. J. Hawksworth, World Economic Forum. http://www.weforum.org/agenda/2018/09/ai-and-robots-could-create-as-many-jobs-as-they-displace. Last accessed 25 Nov 2019
14. The Hays Global Skill Index 2019/20. Hays. Recruiting experts worldwide. http://www.hays-index.com. Last accessed 18 Dec 2019
15. Manyika J.: Job lost, job gained: workforce transitions in a time of automation. Lund, S., Chui, M., Bughin, J., Woetzel, J, Batra, P., Ko, R., Sanghvi, S.: McKinsey Global Institute: http://www.mckinsey.com/~/media/mckinsey/featured%20insights/Future%20of%20 Organizations/What%20the%20future%20of%20work%20will%20mean%20for%20jobs%20skills%20and%20 Jobs-Lost-Jobs-Gained-Report-December-6–2017.ashx. Last accessed 22 Dec 2019

16. A study finds nearly half of jobs are vulnerable to automation. The Economist: http://www.economist.com/graphic-detail/2018/04/24/a-study-finds-nearly-half-of-jobs-are-vulnerable-to-automation. Last accessed 26 Nov 2019
17. Luksha, P., Luksha, K., Peskov, D., Korichin, D.: Atlas novikh professii [The Atlas of New Professions] Moscow school of management "Skolkovo". http://skolkovo.ru/public/media/documents/research/sedec/SKOLKOVO_SEDeC_Atlas.pdf. Last accessed 16 Dec 2019
18. Stylianou, N.: Will a robot take your job? In: Stylianou, N., Nurse, T., Fletcher, G., Fewster, A., Bangay, R., Walton, J. (eds.) BBC: http://www.bbc.com/news/technology-34066941. Last accessed 11 Oct 2019
19. Martin, N., Savchenko, S.: Industria 4.0 Chto budet s rinkom truda? [What expects the labor market]. http://www.dw.com/ru/%D0%B8%D0%BD%D0%B4% D1% 83%D1%81%D1%82%D1%80%D0%B8%D1%8F-40-%D1%87%D1%82%D0%BE-%D0%B1%D1%83%D0%B4%D0%B5%D1%82-%D1%81-D1%80%D1%8B%D0%BD%D0%BA%D0%BE%D0%BC-D1%82%D1%80%D1%83%D0%B4%D0%B0/a-18993560. Last accessed 28 Nov 2019
20. Popkova, E.G., Zmiyak, K.V.: Priorities of training of digital personnel for industry 4.0: social competencies vs technical competencies. On Horiz. **27**(3–4) 138–144 (2019). https://doi.org/10.1108/OTH-08-2019-0058
21. Lysenko, E., Sereda, E.: The new job position in human resources management the talent manager. In: Bekirogullari, Z., Minas, M.Y., Thambusamy, R.X., ICPSIRS 2018—5th International Conference on Political Science, International Relations and Sociology. FUTURE ACAD, pp. 51–58 (2018)
22. Alves, D., Dieguez, T., Conceição, O.: Retaining talents: impact on innovation. In: Paper presented at the Proceedings of the 15th European Conference on Management, Leadership and Governance ECMLG, 36–45 (2019). https://doi.org/10.34190/MLG.19.078
23. McKinsey Global Institute. http://www.mckinsey.com/mgi/overview. Last accessed 12 Dec 2019
24. Curzi, Y., Fabbri, T., Scapolan, A.C., Boscolo, S.: Performance appraisal and innovative behavior in the digital era. Front. Psychol. 10 (2019). https://doi.org/10.3389/fpsyg.2019.01659
25. Bogoviz, A.V., Gulyaeva, T.I., Semenova, E.I., Lobova, S.V.: Transformation changes in the system of professional competences of a modern specialists in the conditions of knowledge economy's formation and the innovational approach to training (2019). https://doi.org/10.1007/978-3-319-94310-7_19
26. Schwab K., Samans R.: The Future of Jobs. World Economic Forum. http://reports.weforum.org/future-of-jobs-2016. Last accessed 10 Nov 2019
27. Coulthard, T.: Annual Manufacturing Report 2019: Taking the pulse of British industry. Coulthard, T. The Manufacturer. http://www.themanufacturer.com/articles/annual-manufacturing-report-2019/. Last accessed 14 Nov 2019.
28. Kaznacheeva, S.N., Chelnokova, E.A., Emelyanova, A.M., Sulimova, I.D.: Stimulirovanie talantlivikh sotrudnikov v organizatsii [Stimulating talented employees in a company]. Int. J. Pract. Theor. Stud. **3**, 126–130 (2018)
29. Shahanipour, S., Amindoust, A., Sahraian, K., Beiranvand, S.: Identification and prioritization of human resource strategies with employees' creativity approach in administrative organizations using SWOT–ANP. Opsearch **57**(1), 119–143 (2020). https://doi.org/10.1007/s12597-019-00416-w
30. Abood, A. Quilligan // Accenture: http://www.accenture.com/ca-en/insight-industry-digital-reinvention?c=ca_ca_industryxo_10394284&n=psgs_generic_1018&g clid=EAIaIQobChMIouSs3eKB4AIVWZ7ACh3QWwWIEAAYAiAAEgLqDPD_BwE. Last accessed 21 Nov 2019
31. Baykal, E.: Digitalization of human resources: E-HR. Tools and techniques for implementing international E-trading tactics for competitive advantage, pp. 268–286 (2019). https://doi.org/10.4018/978-1-7998-0035-4.ch013

32. Manyika J.: Technology, jobs, and the future of work. Manyika, J., McKinsey Global Institute: http://www.mckinsey.com/featured-insights/employment-and-growth/technology-jobs-and-the-future-of-work. Last accessed 12 Nov 2019
33. Bonekamp, L.: Consequences of Industry 4.0 on Human Labour and Work Organisation. In: Bonekamp, L., Sure, M., Journal of Business and Media Psychology. http://journal-bmp.de/2015/12/auswirkungen-von-industrie-4-0-auf-menschliche-arbeit-und-arbeitsorganisation/?lang=en. Last accessed 20 Nov 2019
34. Brandi, J., Kozica, A., Pernkopf, K., Schneider, A.: Flexible work practices: analysis from a pragmatist perspective. Hist. Soc. Res. **44**(1), 73–91 (2019). https://doi.org/10.12759/hsr.44.2019.1.73-91
35. Schuler, R.S., MacMillan, I.C.: Gaining competitive advantage through human resource management practices. Hum. Resour. Manage. **23**(3), 241–255 (1984). https://doi.org/10.1002/hrm.3930230304
36. Järvi, K., Khoreva, V.: The role of talent management in strategic renewal. Empl. Relat. **42**(1), 75–89 (2020). https://doi.org/10.1108/ER-02-2018-0064

Formation of Digital Marketing in Targeted Advertising

Marina Skuratova, Nazym Tulebayeva, and Zhuldyz Abdygaparova

Abstract This article discusses targeted advertising. Targeted advertising in our minds has led to the fact that it is between contextual and media advertising. On the one hand, targeted advertising is based on the principles of bid management, bid management, the need to regulate targeting, and SPS models. On the other hand, targeted advertising provides for a wide coverage, a wide range of visions, and an increase in search demand. Targeted advertising in our minds has led to the fact that it is between contextual and display advertising. This article discusses the advantages, disadvantages, and advantages of targeted advertising placement systems, how to place them, what the scale should be, and what the target audience is intended for placement.

Keywords Targeted advertising · Online advertising · Social network · Online advertising · System analysis · Placement system · Digital advertising

1 Introduction

Most modern Internet systems rely on financial support for online advertising. Due to the fact that advertising is in the best interests of the user, many Internet marketers turn to advertising focused on an advertising broker who is responsible for individually delivering ads that meet their desires and interests. Most existing targeted advertising systems need to access users' profiles in order to investigate their features, but they raise serious privacy concerns and do not want users to participate in advertising systems. This document provides is based on the confidentiality of the system structure to promote targeted advertising. In our product range, an advertising broker is positioned between advertisers and users of a targeted ad and offers a specific reward

M. Skuratova
Plekhanov Russian University of Economics, 117997, 36, Stremyanniy, Moscow, Russia

N. Tulebayeva (✉) · Z. Abdygaparova
Al-Farabi Kazakh National University, al-Farabi Ave. 71, 050040 Almaty, Republic of Kazakhstan
e-mail: tulebaeva_nazym@mail.ru

G. Mutanov and A. Serikbekuly (eds.), *Digital Transformation in Sustainable Value Chains and Innovative Infrastructures*, Studies in Systems, Decision and Control 443,
https://doi.org/10.1007/978-3-031-07067-9_11

rate to encourage users to print interesting and useful ads that do not respond to them. When a user clicks on their behavior, they identify themselves based on their potential conflict of interest and privacy, and advertisers pay an advertising broker per click on the ad. Target audience, focused on specific customers. As an advertising tool, targeting allows you to divide a new (target audience) client who is waiting for news and meets expectations thanks to a well-thought-out advertising mechanism using the Internet. Ad targeting allows you to target specific groups of consumers.

For the first time, targeted advertising appeared on social networks. These platforms started using user survey data to provide accurate and fast search results.

Later, new features appeared in Internet marketing. Browsers collect information about users and search for an audience based on this data.

To use this tool correctly, you need to understand the psychology of your target audience, the interests of your clients, and their socio-demographic parameters. Let's look at a simple example to confirm this and understand how targeted advertising works.

2 Literature Review

People find themselves more in social networks. This is done for many purposes, as they play the role of consumers, they search for information about products, buy and consume them, and tell others about their experiences as they see fit. Marketers are responding to fundamental building blocks by expanding the use of digital marketing channels. By 2017, a third of global advertising spending will be delivered to digital channels. Thus, in the future, consumer marketing will be focused on the digital environment, especially on social networks and mobile devices. Therefore, marketing research is necessary to study, understand, and analyze consumer behavior and taste in the digital environment.

The work of scientists Ziyadin S., Doszhan, R., etc. explores how consumers perceive, experience and are influenced by the digital environment in which they are located, as part of their daily life [1].

Ziyadin S. and Serikbek A. also write that one of the new directions of modern marketing in the global web has become the Internet marketing of educational services. They also say that the object of marketing educational services are educational organizations, consumers of educational services, both legal entities and individuals, as well as the state and directly those bodies that determine the rules and conditions for licensing and accreditation of educational organizations, as well as those organizations that participate in the promotion of educational services on the market [2].

In their work, Turow J. and King J. explore the opinions of Americans about behavioral targeting of marketers-a controversial issue currently facing government policy makers. They write that behavioral targeting involves two types of activities: tracking user actions and then adapting ads for users based on these actions [3].

3 Methodology

Like classic video advertising, Internet video is used. Yes, but the airtime on the TV is much cheaper. In addition, online video advertising offers advertisers a number of tools, such as targeting and real statistics (Fig. 1).

Let's look at and analyze what tools you can use to attract target audiences:

- Re-marketing. You can already attract those users who will interact with any interaction that occurred on your video or YouTube channel. If you have linked your YouTube account to AdWords, the list of repeat ads will be created automatically.
- Demographic data. Why pay for an empty-point action. Set targeting for users of the desired age and gender.
- Keyword. Context keys compare videos with web content. It works very well for increasing sales.
- Topics. Everything is easy here: showing a video on a Youtube video about a specific topic.

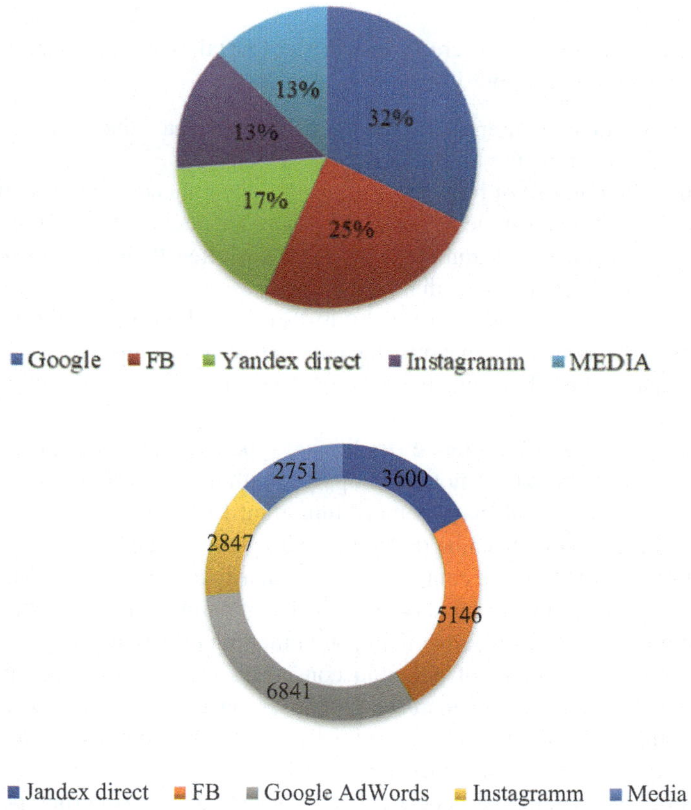

Fig. 1 Distribution of requests by the advertising channel

Table 1 Pros and cons of targeting ads in Instagram

	Plus	Minus
1	Low entry threshold in contrast to other advertising methods	Complexity of settings/Cabinet (at first glance)
2	Accurate targeting of your target audience	There is no trust in advertising
3	Speed	You can't reach your audience without Facebook
4	Consistency of results	Blocking payment cards and advertising cabinets
5	Scalability	Dear subscribers. Relative to other methods
6	The reach of advertising to the user	There is no clear need for the product or service you are selling
7	Clear measurable results	
8	Analytics. When working with the site	
9	Ability to advertise for an active audience in your profile	

- Thematic audience. You can create an audience that takes into account the specifics of your business [4] (Table 1).

We have formulated an approximate list of various goals that can be set for an advertising campaign on the Internet.

Branding: the purpose of branding campaigns is to enhance various components of the brand, such as awareness, image, or customer intent. These campaigns are usually part of a broader media strategy, and most often their combined effect is evaluated together with other media.

The work to increase customer loyalty is also carried out within the framework of branding and is aimed at expanding the experience of the consumer's relationship with the brand online, but with the specific goal of interacting directly with the Internet user [5].

Growth of sales in offline mode (traditional types of sales) continues to be the ultimate goal of the brand promotion campaign, with the exception of image advertising. Achieving this goal may be more important for specific campaigns, which means that the assessment can be performed with greater accuracy.

Online sales growth: these campaigns are primarily aimed at increasing sales on the brand's website or in partner networks. Such campaigns can also simultaneously pursue the goals of achieving advertising and branding effectiveness.

Determining the number of potential consumers is based on a result-oriented method, but does not aim to increase purchases on the Internet. The goal is to study the range of people who use interactive Internet resources, such as games and questionnaires, for classification purposes [6].

Increasing the reach and repeatability of messages: as Internet penetration increases, it quickly becomes a medium that allows you to increase the impact of an advertising message. One of the goals of Internet campaigns may be to expand the impact of the message and ensure repeatability in other media.

Reducing the cost of attracting customers may be another goal of online campaigns. For result-oriented advertisers, it is the main one.

Different indicators can be used to achieve each of the above goals. The following is a basic but not exhaustive list that provides an overview of the evaluation tools available to advertisers. In many cases, the same indicator can be used to measure the achievement of different goals [7] (Table 2).

Advertising based on the use of Internet users by sending, the definition of which involves user actions taking into account the target audience, leads to the fact that the previous actions of users are not on the Internet, but threats that caused harm to the result (registration, receipt, visit, site that did not complete the introduction to the product). Many online retailers use retargeting, because, according to Criteo, companies that specialize in determining the target audience based on user actions, accumulating information about them, 90% of visitors leave the site without making a purchase. Taking into account that 70% of visitors are directed to e-Commerce sites by a link to paid advertising, we can say that thanks to this method, the conversion rate or index of the number of site visitors increases dramatically [8]. But targeted advertising has its drawbacks:

Table 2 9 types of indicators for evaluating the effectiveness of online activity

Impressions (display) • number of impressions • number of" visible " ads • number of ads viewed • the timing of	Conversion • coefficient of CTR • conversion rate • conversion rate after viewing • click-through conversion rate	Traffic • number of visits • number of pages viewed • duration of the visit • number of rejections from visits
Interaction • coefficient of the interaction • interaction time • indicator of viewed videos • indicator of fully viewed videos • video viewing time • social media activity	Subscription • number of subscriptions (requests for information, games, brochures, etc.) • the subscription price • recommendation coefficient (social networks, "viral" online marketing campaign on the Internet)	Media • General rating point (GRP) • memorability of ads • the level of coverage • the rate of repeat messages
Sales (distribution) • additional sales • revenue received • conversion rate for the target audience • influence on the frequency and volume of purchases • traffic at retail outlets initiated by advertising on the Internet	Return on investment (ROI) • cost of attracting buyers (offline versus online) • return on investment (revenue received/ advertising expenses)	Post-tests • impact on recognition • the impact on memorability of advertising • influence on the brand image • influence on purchasing intentions • influence on recommendations

- A small level of motivation to go to the link •. Because the person who saw this ad Did not want to buy something, they entered the social network.These ads may be the reason that the consumer most often tries to open.
- For this reason, the conversion rate is much lower.
- It is not possible to advertise certain types of goods and services (alcohol, tobacco, or products that cause harm to health) that are prohibited by the moderator
- Unacceptable to all [9].

4 Results and Analysis

Digital media is expected to account for more than 51% or $ 240 billion of planned marketing spending in 2019.Currently, search advertising is the most popular type of digital advertising: marketers 2018 increase advertising spending by 12%. Therefore, this year, 67% of all search ads will be mobile, and the total cost of this category of online advertising will amount to 29.5 billion dollars [10].

However, research on this topic provides information in which research and analysis of factors that may affect the effectiveness of Facebook ads, limited or special and other researchers have been conducted. Previous research has analyzed and placed ads on the Facebook platform for a company that is a traditional food manufacturer, and found that the most profitable ads are mobile ads. It is one of the most effective methods in terms of cost and time. Surprisingly, none of them believe that they give a good return on investment in Facebook advertising (ROI). Facebook claims to offer the cheapest rate for ads that meet their rules, requirements, and regulations and attract more of their users. In other words, the cost of advertising with a high coefficient of attraction will be lower. How consumers perceive marketing communications depends on the relevance of advertising and the ability to introduce marketing communications into the consumer's mind. Which types of social media ads significantly affect the relationships of their members are related to the disruptive advertising section, as well as the parts covered by the potential development model (ELM). Which features of advertising in social networks should be more attractive, and which ones are annoying, should be created in such a way that the consumer can remain conscious. Because if necessary, your ad should be a reminder of the first one. This is handled both in the ad and in the entire model. These features only affect ads if the buyer can receive marketing messages, so the ability to receive is a feature in the mode. If a consumer can get a marketing relationship, he/she can form an attitude towards it, it is divided into negative and positive relationships that form the level of engagement. This also shows that the level of engagement affects decision—making [11].

The mobile video demo will also change. For example, Snapchat outputs a feature that users can swipe photos of friends left and right with their finger for editorial content where marketers can participate and suggest. This is due to the fact that in the new year, income growth is 2.5%, despite the fact that there was a fall. This corresponds to the sum of 183 billion dollars. Annual revenue from advertising on

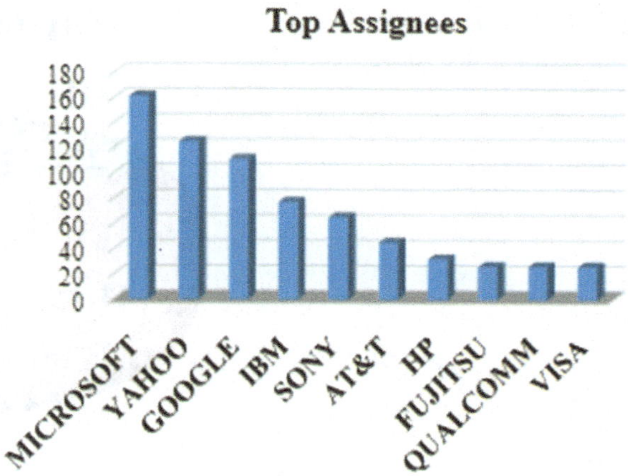

Fig. 2 Top 10 players patented in the field of web advertising

the Facebook social platform is expected to grow by 32.1%, while Google advertisers spend 14.8% more. In addition, Facebook will test the ad display feature before all native videos, which will help marketers expand access to the most popular sites in social networks today [12].

The most important thing is that online advertising does not like the Asian games: according to the latest data, 90% of mobile devices that block ads operate in Asia.

If nothing is done to solve the problem of ad blocking, this policy will exceed $ 75 billion for e-Commerce sites by 2020.

Figure 2 shows the top 10 players patented in the field of web advertising that are regulated. The maximum number of patents in this area is about 161 Microsoft, followed by Yahoo, Google and IBM. There are also very few patents of many companies/individuals. About two-thirds of the concessionaries are united by a single patent organization [13].

Figure 3 shows the various criteria used for targeting ads, such as previous browsing history, user Profile, events related to vehicle mobility and storage locations, and so on.Analysis has shown that ads based on a user's profile are the preferred criteria that users follow closely in their browsing history [14].

Figure 4 above shows trends based on demographic categories. Targeting based on location derived from demographic information is most preferred, since high patent activity is visible in this area. The chart 6 above also shows that with the advent of social media, the use of social media for ad targeting has appeared since 2018 and activity in this area has been slowly increasing in recent years [15, 16].

Digital advertising is a major topic in the marketing literature, and as far as consumer behavior is concerned, it looks at how consumers respond to various aspects of digital advertising. Schumann considered how to overcome negative reactions to personalization with appeals to normative reciprocity (instead of appeals to utility). Lambrecht studied ad retargeting, where personalized recommendations based on previous browsing histories are made when the consumer returns to the site. Negative reactions to retargeting occur, but they are mitigated when consumer preferences

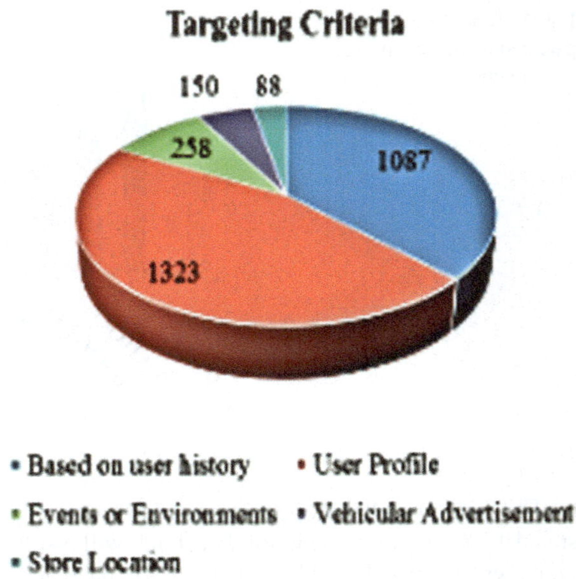

Fig. 3 The criteria used for targeting advertisements

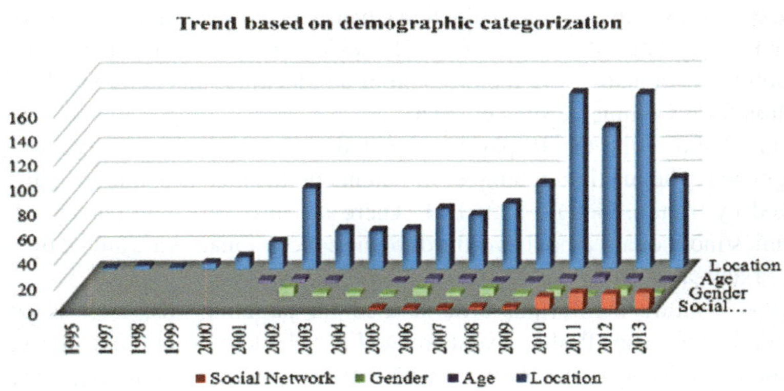

Fig. 4 Trends based on demographic categories

become more accurate. He found that personalized advertising on a website is more favorably perceived when consumers better perceive control over personal information used for personalization, which is directly consistent with the literature on psychological responses and offers a theoretical way forward [17].

5 Conclusion

In this work, we studied the problem of targeted advertising with regard to privacy, and also proposed a compensation scheme to encourage users to watch interesting ads. The compensation structure is aimed at promoting targeted advertising by creating irrational conditions. Within this framework, we analyze the interaction between advertisers, the advertising broker, and users through step-by-step game modeling. The results of the simulation show the effectiveness of the proposed structure, in which users are motivated to suppress interested ads, and the advertiser and advertising broker achieve significant success in their revenue. Thus, with the appropriate design, the compensation system can accept targeted ads and bring profit to all organizations [18].

References

1. Ziyadin, S., Doszhan, R., Borodin, A., Omarova, A., Ilyas, A.: The role of social media marketing in consumer behaviour. E3S Web Conf. **135**, 04022. EDP Sciences (2019)
2. Ziyadin, S., Serikbek, A.: Internet-marketing in the sphere of higher education. Lect. Notes Netw. Syst. **84**, 290–297 (2020)
3. Turow, J., King, J., Hoofnagle, C.J., Bleakley, A., Hennessy, M.: Americans reject tailored advertising and three activities that enable it. SSRN 1478214 (2009)
4. Toubiana, V., Narayanan, A., Boneh, D., Nissenbaum, H., Barocas, S.: Adnostic: Privacy preserving targeted advertising. In: Proceedings Network and Distributed System Symposium (2010)
5. Guha, S., Cheng, B., Francis, P.: Privad: Practical privacy in online advertising. In: USENIX Conference on Networked Systems Design and Implementation, pp. 169–182 (2011)
6. Kodialam, M., Lakshman, T.V., Mukherjee, S.: Effective ad targeting with concealed profiles. In: Proceedings IEEE INFOCOM, pp. 2237–2245. IEEE (2012)
7. Dave, V., Guha, S., Zhang, Y.: Measuring and fingerprinting click-spam in ad networks. In: Proceedings of the ACM SIGCOMM 2012 Conference on Applications, Technologies, Architectures, and Protocols for Computer Communication, pp. 175–186 (2012)
8. Yan, J., Liu, N., Wang, G., Zhang, W., Jiang, Y., Chen, Z.: How much can behavioral targeting help online advertising? In: Proceedings of the 18th International Conference on World Wide Web, pp. 261–270 (2009)
9. Arthur, L.: Big Data Marketing: Engage your Customers more Effectively and Drive Value. John Wiley & Sons (2013)
10. Bakshy, E., Hofman, J.M., Mason, W.A., Watts, D.J.: Everyone's an influencer: quantifying influence on twitter. In: Proceedings of the Fourth ACM International Conference on Web Search and Data Mining, pp. 65–74 (2011)
11. Tovma, N.: External and internal factors affecting the development of the digital economy. In: Proceedings of the 32nd International Business Information Management Association Conference, IBIMA 2018 - Vision 2020: Sustainable Economic Development and Application of Innovation Management from Regional expansion to Global Growth, pp. 5899–5902 (2018)
12. Kim, D.Y., Kim, H.Y.: Influencer advertising on social media: the multiple inference model on influencer-product congruence and sponsorship disclosure. J. Bus. Res. (2020)
13. Shatnawi, M., Mohamed, N.: Statistical techniques for online personalized advertising: a survey [C]. In: ACM Symposium on Applied Computing, ACM, pp. 680–687 (2012)

14. Schumann, J.H., von Wangenheim, F., Groene, N.: Targeted online advertising: using reciprocity appeals to increase acceptance among users of free web services. J. Mark. **78**, 59–75 (2014)
15. Lambrecht, A., Tucker, C.: When does retargeting work? Information specificity in online advertising. J. Mark. Res. **50**, 561–576 (2013)
16. Tovma, N.A.,Shurenov, N.B., Bimendiyeva, L.A., Kozhamkulova, Z.T., Akhmetova, Z.B.: Territorial marketing and its role in determining regional competitiveness. Evaluating supply chain management. Uncertain Supply Chain Manage. **8**, 1–16 (2019)
17. Danaher, P.J., Mullarkey, G.W.: Factors affecting online advertising recall: a study of students. J. Advert. Res. **43**, 252–267 (2003)
18. McCoy, S., Everard, A., Polak, P., Galletta, D.F.: The effects of online advertising. Commun. ACM **50**, 84–88 (2007)

Digitalization as a Driver for the Development of Strong Regions and Urbanization

Recursive Optimization Model for Constructing a Hamilton Tree (G-tree) of Minimum Cost not Belonging to the Class NP

Serik Ospanov and Saltanat Kondybayeva

Abstract This article deals with the traveling salesman problem (TSP). For TSP belonging to the class of NP-completeness, the existence of an efficient algorithm with estimated complexity measures is not proven. The considered model of the traveling salesman problem (TSP) not belonging to the NP class is formed as a single-source problem for recursively constructing paths of Hamilton trees (G-tree) emanating from a fixed *n*-dimensional closed graph node and branched along subtrees of length *n-1*. The structure of the recursive model is built from a subset of the paths formed from a sequence of stationary & admissible edges on a given node. Recursive costs are placed on the allowed segments nested in each other in a strictly descending order according to the ordinal numbers of the constructed G-trees. The monotonously descending series of positive numbers formed in the limit defines the Minimum-Cost Spanning G-tree. To construct G-trees by an iterative algorithm, without checking for belonging to the class NP, a recursive model is formulated in the computational aspect. In this case, an individual combinatorial task at the computer input is represented using two recursive algorithms. Two recursive algorithms and the value of the initial G-tree represent an individual combinatorial task.

Keywords Recursive optimization model · Traveling salesman problem · NP class · Hamilton trees · Optimality criterion

1 Introduction

The Traveling Salesman Problem (TSP) is one of the most famous problems in discrete mathematics. The traveling salesman problem (TSP) is used in economics in the design of transport and logistics systems and for choosing the route of minimum cost from a variety of options. The problem statement is related to the search of the minimum weight in the full weighted graph G of the Hamiltonian cycle (Traveling

S. Ospanov · S. Kondybayeva (✉)
Al-Farabi Kazakh National University, al-Farabi Ave. 71, 050040 Almaty, Republic of Kazakhstan
e-mail: Saltanat.Kondybaeva@kaznu.kz

G. Mutanov and A. Serikbekuly (eds.), *Digital Transformation in Sustainable Value Chains and Innovative Infrastructures*, Studies in Systems, Decision and Control 443,
https://doi.org/10.1007/978-3-031-07067-9_12

Salesman Path). The Traveling Salesman Problem has a special place in combinatorial optimization and operations research and is one of the typical NP-hard problems.

Traveling salesman models mathematically describe the tasks of finding a rational option for transporting goods while visiting all the intermediate points [1–4]. Any company engaged in transportation is financially interested in determining the least cost route [5].

The traveling salesman problem (TSP) is studied in works on the graph theory, as well as in combinatorial and computational optimization in various statements [6–10]. The optimization statement is the problem of discrete programming with Boolean variables [11, 12].

On a symmetric closure graph of the n-order, TSP is formulated as follows: a Hamiltonian circuit is constructed in such a way that the total weight (cost) of the arcs forming it is minimal [13].

2 Literature Review

The Traveling Salesman Problem drew the attention of scientists from RAND Corporation research center in the 50s and 60s of the last century. J. Danzig, R. Falkerson and S. Johnson formulated the problem as a discrete optimization problem on graphs for the first time and applied the branch and bound method (Danzig et al. 1954) to solve it.

The authors constructed a Traveling Salesman Path for statement of the problem with 49 cities and substantiated its optimality but their result had a private concernment. In fact, the branch-and-bound algorithm is a variation of exhaustive search with a reduction in the subsets of feasible solutions that obviously do not contain optimal routes. Approaches were developed in many papers to find exact solutions for linear integer programming problems that made it possible to construct optimal paths for higher dimensional statements.

M. Held and R. Karp proposed an exact algorithm for the general statement of the traveling salesman problem at the beginning of the 1960s [14], based on the dynamic programming method [15] with O ($n^2 2^n$) complexity. It is difficult to overestimate the importance of the Held-Karp algorithm since for almost fifty years it has been the best accurate method for solving the general formulation of the problem. The next step in this direction was the paper by A. Bjorklund, published in 2010 [16]. The author was able to construct an exact algorithm for the statements of the traveling salesman problem over graphs with n state points and positive edge weights which has O (1.657^n) complexity. A survey of exact algorithms for the traveling salesman problem and other intractable problems of combinatorial optimization problems, as well as a list of open questions in this area are given in G. Weeginger's papers [17, 18]. Undoubtedly, exact algorithms for TSP are important from theoretical grounds, however, the exponential time from the number of state point of a given graph makes it difficult in general to use them for practical purposes.

Tucker et al. [19] formulated a discrete programming model that reflects the conditions of the traveling salesman problem with the least number of constraints and independent binary variables. To determine the structure of the Tucker et al. [19] optimization model, the following matrix with binary elements $X = [x_{ij}]_{(n+1)(n+1)}$, where $x_{ij} = 1$, is preliminary added in cases where the Hamiltonian circuit has an arc (i, j) [10].

Then the TSP model takes the following form:

$$f(x) = \sum_{i=0}^{n}\sum_{j=0}^{n} C_{ij}x_{ij} \tag{1}$$

$$\sum_{i=0}^{n} x_{ij} = 1,\ j = 1, 2, \ldots, n \tag{2}$$

$$\sum_{j=0}^{n} x_{ij} = 1,\ i = 1, 2, \ldots, n \tag{3}$$

$$u_i - u_j + nx_{ij} \le n - 1,\ 1 \le i \ne j \le n \tag{4}$$

$$u_i \in R\text{-real numbers},\ i = 1, 2, \ldots, n \tag{5}$$

Any solution satisfying the conditions (1)–(3) and representing a Hamiltonian circuit also satisfies condition (4). The opposite is equally true. In the case when condition (4) is not satisfied, the solution has at least two elementary contours with the number of arcs $k \le n$.

In the Tucker et al. [19] model, it is required to show that for each Hamiltonian circuit originating at the vertex i_0, it is possible to find a value u_j that would satisfy condition (4). Suppose $u_j = r$ in the case when the vertex i is Hamiltonian for the arc r when passing the path from i_0 to i, where $i = 1, 2, \ldots, n$.

It is clear that for $x_{ij} = 0$, the condition $u_i - u_j \le n - 1$ is satisfied for each arc (i, j). Therefore, it is performed for all $x_{ij} = 0$.

For $x_{ij} = 1$ we get

$$u_i - u_j + nx_{ij} = r - (r + 1) + n = n - 1 \tag{6}$$

Using this modeling method, in the problem represented by a graph with n vertices, the number of binary variables becomes n^2 (instead of n^3 variables), and the number of constraints is $2(n - 1) + (n - 1)^2 = n^2 - 1$ (instead of $2n^2$).

Thus, for $n = 6$, we get a model with $36 = n^2$ binary variables and $2(n - 1) + (n - 1)^2 = n^2 - 1 = 36 - 1 = 35$ constraints.

However, due to the square dependence of the number of constraints and the number of variables on the number of vertices, these numbers grow very rapidly with increasing n.

The structure of the Tucker et al. [19] model of TSP includes the basics of the problems of discrete optimization, graph theory and pattern recognition [20, 21].

A number of results related to the study of the computational complexity of combinatorial optimization problems arising in such areas as graph theory, mathematical programming, combinatorial optimization, mathematical logic, etc. were obtained in the 70s. P and NP classes were introduced for the first time in the fundamental work of Cook [22] in 1971, and the NP-completeness of Boolean satisfiability problem was proved. R. Karp substantiated NP-completeness for 21 problems including the Hamiltonian cycle problem a year later by reduction of satisfiability problem of Boolean forms to it [23]. The number of known NP-complete problems grew rapidly [24] after the publication of this paper. S. Sahni and T. Gonzalez substantiated the NP-difficulty to construct the approximate solution with O (2^n) accuracy for TSP general statement in 1976 [25].

H. Papadimitriou showed at the end of the 70s that even the Euclidean statement of the traveling salesman problem is NP-hard [26]. Thus, the exact solution of the Euclidean statement of the problem and the approximate solution with a guaranteed estimate of the accuracy of the general statement of the problem cannot be found in polynomial time under the assumption $P \neq NP$. The matter to construct efficient approximate algorithms for TSP subclasses arose naturally.

3 Methods and Materials

The traveling salesman problem is located at the intersection of continuous and combinatorial optimization problems. To solve the TSP, various heuristic algorithms of the branch and bound method have been developed [26, 27]. However, the importance of effective lower bounds in the branch and bound method leads to large search trees [28–30].

In the individual traveling salesman problem, a cost matrix is set between all city pairs $C^o = \left[C^o_{ij}\right]_{n \times n}, i \in I^o, j \in J^o$ [30, 31].

The problem is to find a detour with a minimum total cost. Here, a detour is a closed route that passes through each city exactly once [2–4, 32].

In the modern theory of building and analyzing algorithms, the complexity of the optimization problem solution is characterized by its belonging to the NP class—complete problems [33–35]. Effective computational algorithms have not yet been found to solve problems of this class.

Problems that are solved during the polynomial time $O(n^k)$, where n is the size of the input data and k is a constant belong to the category of easily solvable [36, 37]. Problems with times exceeding the polynomials are accepted as difficult to solve or more complex. However, algorithms with polynomial operation time have not yet been developed to solve NP-complete problems.

NP completeness [22] applies directly to decision-making (decision) problems, where each valid solution can be associated with the value of the objective function.

At the same time, no theoretical proof exists that there is no effective algorithm for problems from the NP-class that have real applications [24, 38, 39]. The traveling salesman problem, which is solved to make optimal infrastructure decisions, belongs to the NP-class of complete tasks.

This paper represents a recursive-decomposition (RD) model for constructing the stationary allowable contours of a traveling salesman problem. To solve it, an iterative algorithm has been developed for finding the optimal solution of the minimum cost that works without looping. Based on the properties of recursive relations, the computational efficiency of the iterative approach is proved.

In a single-source RD model, the process of building G-trees [40] of length $(n - 1)$ starts from the base arcs $C^0M(i_0, j_0) = \{(i_0, j_0), j_0 \in J^0_{i_0}\}$ and branch in $(n - 1)$ directions. In each direction, the paths of G-trees are constructed in the form of a sequence of recursive edges of the minimum—admissible (EMA) costs.

The structure of the recursive model, consisting of the preliminary and iterative stages, provides systematic construction of path segments according to the recursive—iterative scheme.

At the preliminary stage (nodes i_0 and i_1) of the model functioning, the following recursive data and relationships are formed for building trees in the directions.

(1) Forming a reduced matrix $C^0R = \left[C^{0r}_{ij}\right]_{n\times n}, i \in I^0, j \in J^0$, obtained from the cost matrix $C^0R = \left[C^{0r}_{ij}\right]_{n\times n}, i \in I^0, j \in J^0$ using the following reduced coefficients

$$C^{0r}_{ij} = \left(C^0_{ij} - \alpha^0_i\right) - \beta^0_j, i \in I^0, j \in J^0;$$

$$\alpha^0_i = \min_{j\in J^0} C^0_{ij}, i \in I^0;$$

$$\beta^0_j = \min_{i\in I^0}\left(C^0_{ij} - \alpha^0_i\right), j \in J^0$$

Denoting the sum of the reduced coefficients:

$$\varphi^0 = \alpha^0 + \beta^0 = \sum_{i\in I^0} \alpha^0_i + \sum_{j\in J^0} \beta^0_j;$$

(2) Fixing from the graph $G(C^0)$ a node as a source for constructing G-trees;
(3) Building the initial contour of the salesman coming from $i_0 \in I^0$ and choosing its value $f_{i_0,u=0} = f(x^n_{i_0,u=0})$.
(4) Building the initial segment of admissible costs of G-trees originating from $i_0 \in I^0$; $\Delta\left(f_{i_0,u=0}\right) = f_{i_0,u=0} - \varphi^0_{u=0}$.
(5) Forming C^0 and C^0R matrix with constraints:

(a) $C^0 = \left[C^0_{ij}\right]_{n\times n}$

$$C_{ij}^{0} \leq \Delta_0\left(f_{i_0,u=0}\right) = f_{i_0,u=0}, i \in I^0, j \in J^0;$$

(b) $\dot{C}^0 R = \left[C_{ij}^{0r}\right]_{n \times n}$

$$C_{ij}^{0r} \leq \Delta_0\left(\varphi_{u=0}^{0}\right) = \Delta_0\left(f_{i_0,u=0}\right) - \varphi^0, i \in I^0, j \in J^0;$$

At the iterative stage of the recursive model, the process of constructing the G-tree in the selected direction continues as per the iterative scheme at the nodes *V.3* ÷ *V.(n* − 1).

At iterative $V \in \{V.3 \div V.(n-1)\}$, the minimum admissible edges $(i_v, j_v) \in i_v \times J_{i_0}^{v}$ extends the segment of the G-tree built on the $(V-1)$ step:

$$C^{v-1} M(i_0, j_0) = \left\{(i_v, j_k)_{k=0}^{v-1}\right.$$

If there is no edge of the minimum admissible cost at the next iterative step, the constructed segment is reduced to the first subtree with a nonempty EMA. Then the iterative algorithm continues the process of constructing the G-tree in the chosen direction.

The optimality criterion of the recursive model checks the constructed G-trees and determines the Minimum-Cost Spanning G-tree.

Using the reduced submatrix $C^{v-1} R(i_{v-2}, j_{v-2})$, the following algorithm of three recursive operations is used to determine the cost functions of C^0 and C^{vr} for the minimum admissible edge:

V.0—the operation of deleting the main element (i_{v-1}, j_{v-1}) and the formation the submatrix of *(n-v)* order with restrictions:

$$C^v(i_{v-1}, j_{v-1}) = \left[C_{ij}^{v}\right]_{(n-v)(n-v)}$$

$$C_{ij}^{v} \leq \Delta_v\left(f_{i_0,u}\right) = \Delta_{v-1}\left(\varphi^{v-1}\right) - \varphi^{v=1}(\alpha_{i_{v-1}, j_{v-1}})$$

V.0.0—the formation of a reduced submatrix with restrictions:

$$C^v R(i_{v-1}, j_{v-1}) = \left[C_{ij}^{vr}\right]_{(n-v)(n-v)}, \varphi^v = \alpha^v + \beta^v$$

$$C_{ij}^{vr} \leq \Delta_v(\varphi^v) = \Delta_v\left(f_{i_0,u}\right) - \varphi^v$$

V.0.1—the operation of forming a subset of index-arcs to determine the minimum admissible edges $(i_v j_v)$, the operation is performed in the following sequence:

(1) Formation of a subset of index-arcs
$I(C_{i_v}^{vr}) = \{(i_v, j) \in i_v \times J_{i_0}^{v} : C_{i_v j}^{vr} \leq \Delta(\varphi^v), j \in J_{i_0}^{v}\};$

(2) Highlighting rows

$$C^0 = \left(\left(C_j^0, j \in J\right)\left(C_{i_v}^{vr}\right)\right), C_{i_v}^v = (C_{i_v j}^{vr}, j \in I(C_{i_v}^{vr}));$$

(3) Definition of Minimum Cost Element

$$C_{i_v j_v}^0 = min\{C_{i_v j}, j \in I(C_{i_v}^{vr})\}.$$

Selection of index-arc $(i_v, j_v) \in i_v \times J_{i_0}^v$ and corresponding element $C_{i_v j_v}^{vr}$ from a reduced row $C_{i_v}^{vr} = (C_{i_v j}^{vr}, j \in J_{i_0}^v)$.

The base interval $\Delta_0\left(f_{i_0,u=0}\right) = f_{i_0,u=0} - \varphi_{u=0}^0$ constructed at the preliminary stage of the recursive model limits the upper and lower boundaries of the costs of all minimum admissible (MA) G-trees originating from $i_0 \in I^0$ and cost limited by the value $f_{i_0,u=0} = f(x_{i_0}^n)$.

At the iterative stage of the recursive model are iteratively and consistently constructed (MA) G -trees $C_u^{n-1}(i_0, j_0) = \left\{(i_v, j_v)_{v=0}^{n-1}\right\}_u$, $u = 1, 2, \ldots, m$, where $m < \Delta_0(f_{i_0})$ with direct objective function $f_{i_0,u-1} > f_{i_0,u}$, $u = 1, 2, \ldots, m$ and recursive objective function $\varphi_u^0 \le \varphi_{u+1}$, $u = 1, 2, \ldots, m$ values.

For traveling salesman problems (TSP) belonging to the NP-class, there are no efficient algorithms with evaluative measures of polynomial complexity [6]. The recursive optimization (RO) model for constructing (MA) G-trees does not belong to class N and is solved without resorting to a full enumeration algorithm.

The pseudopolynomial iterative algorithm developed to solve the RO model constructs a subset of MDG trees in the order of strictly monotonic descent in the values of the objective functions $f_{i_0,u-1} > f_{i_0,u}$ $u = 1, 2, \ldots, m$.

The pseudopolynomial iteration algorithm works without internal looping [10, 36, 37, 41]. The pseudopolynomial algorithm solves any recursive problem in a time limited by a polynomial in the length of the recursive codes $n - v$, $v = 0, 1, \ldots, n - 1$ and the largest value of the positive numbers involved in the iterative process is equal to $f_{i_0} = f(x_{i_0}^n)$.

In the G-tree structure, each edge of the minimum admissible cost is measured simultaneously by two recursive measurements: the length of the code and the cost of the edges. Based on these measurement features, the computational complexity of the pseudopolynomial iterative algorithm is estimated by the additive function of two arguments.

The first additive function estimates the computational complexity of the algorithm by the length of the code. This estimate is polynomial.

The second additive function, bounded above by the recursive value $f_{i_0,u=0} = f(x_{i_0,u=0}^n)$, estimates the number of (MA) G-trees required to build a (MA)G_u tree of minimal cost.

4 Results

4.1 *Specifics of the Formation of a Recursive-Decomposition Model of the Traveling Salesman Problem not Belonging to NP*

New construction of (MA) G-tree.

Set: Cost Matrix

$C^0 = \left[C_{ij}^0\right]_{n \times n}, i \in I^0, j \in J^0;$

Choose a node $i \in I^0$ and determine the initial contour of the traveling salesman problem $C^{n-1}M(i_0, j_0) = \left\{(i_v, j_v)_{v=0}^{n-1}\right\}$ with the cost $f_{i_0} = f(x_{i_0}^n)$.

Construct a dimensional full closed graph of an individual problem:

$$\begin{aligned}
&G\left(C^0\right) = \left\{\left(V, E, C^0, f_{i_0}\right) : V = (1, 2, \ldots, n);\right.\\
&C^0 = \left(C_{ij}^0, j \in J^0\right), i \in I^0;\\
&C^0 = \left[C_i^0, i \in I^0\right];\\
&e_i = \left\{(i, j) \in i \times J^0 : C_{ij}^0 \in C_i^0\right\};\\
&\left.E = \left[e_i, i \in I^0\right] = [i, j]_{n \times n}\right\}.
\end{aligned}$$

On the graph $G(C^0)$, the matrix C^0 is represented by rows. The set of edges of the graph E is represented as index pairs of the matrix C^0elements.

In the recursive model, the process of constructing the traveling salesman contour of length $n - 1$, belonging to the graph $G(C^0)$, begins with choosing the root node $i_0 \in I^0$ and subsets of guide arcs $C^0M(i_0, j_0) = \{(i_0, j_0)\}, j \in J_{i_0}^0$.

We denote the set of salesman contours of the graph $G(C^0)$.

$GM\left(C^0\right) = \left\{C^{n-1}M(i_0, j_0) = \left\{(i_v, j_v)_{v=0}^{n-1};\right.\right.$

$\left.i_v = j_{v-1}, v = 1, 2, \ldots, n-1; j_{n-1} = i_0\right\} \subset G(C^0), \left|GM(C^0)\right| = (n-1)!$. This set is decomposed into mutually disjoint subsets along the directions of arcs $C^0M(i_0, j_0), (i_0, j_0) \in i_0 \times J_{i_0}^0$:

$$\begin{aligned}
&GM_{j_0}\left(C^0\right) = \left\{C^{n-1}M_{j_0}(i_0, j_0)\right\}, j_0 \in J_{i_0}^0,\\
&\left|GM_{j_0}\left(C^0\right)\right| = (n-2)!;\\
&GM\left(C^0\right) = \bigcup_{j_0 \in J_{i_0}^0} GM_{j_0}\left(C^0\right)
\end{aligned}$$

4.2 Algorithm of Formulation a Recursive Model for Constructing an Optimal G-tree

Required to construct the (MA) G-tree

$$C^{n-1}M(i_0, j_0) = \left\{(i_v, j_v)_{v=0}^{n-1}\right\}_{j_0 \in J_{i_0}^0}$$

belonging to subsets $GM_{j_0}(C^0) \subset GM(G^0)$, $j_0 \in I_{i_0}^0$, where $GM_{j_0}(C^0) \neq GM_{j_0^2}(C^0)$, $j_0^1 \neq j_0^2$ and delivering a pair of double recursive functions $f(X_{i_0}^n) = \sum_{v=0}^{n-1} C_{i_v j_v}^0 x_{i_v j_v}^{v+1} = \sum_{v=0}^{n-1} C_{i_v j_v}^0 \to \varphi^{n-1}(\alpha_{i_{n-1}, j_{n-1}}) = \sum_{v=0}^{n-1} \varphi(\alpha_{i_v, j_v})$ where $x_{i_v j_v}^{v+1} = 1, v = 0, 1, \ldots, n-1; i_v = j_{v-1}, v = 0, 1, \ldots, n-1; j_{n-1} = i^0$ minimum value belonging to the base range of admissible costs $\Delta_0(f_{i_0, u=0}) = f_{i_0, u=0} - \varphi_{u=0}^0$.

The model is an (n – 1)-level system of recursive relations designed to build traveling salesman contour trees coming from the vertex $i_0 \in I^0$ of the graph $G(C^0)$ in the directions $C^0 M(i_0, j_0^t) = \{(i_0, j_0^t)\}$, $t = 1, 2, \ldots, n-1$. The model consists of an initial and iterative stage, respectively, with node numbers $V = 0, 1$ и $V = 1, 2, 3, \ldots, n-1$.

At the preliminary stage, recursive relationships are formed that allow from the basic guide arcs $C^0 M(i_{0,} j_0^t)$, $t = 1,2, \ldots, n-1$ constructing at an iterative stage the travelling salesman problem contours

$$C_u^{n-1} M(i_{0,} j_0^t) = \left\{(i_v, j_v)_{v=0}^{n-1}\right\}, u = 1, 2, \ldots, m_t, t \in T(f_{0,0}).$$

Suppose that a fixed initial traveling salesman contour is known $C^{n-1}M\left(i_0, j_0^{\bar{t}}\right) = \left\{(i_v, j_v)_{v=0}^{n-1}\right\}$, $\bar{t} \in T$ with cost

$$f_{\bar{t}} = f\left(x_{\bar{t}}^n\right) = \sum_{v=0}^{n-1} C_{i_v j_v}^{0,\bar{t}} X_{i_v j_v}^{v+1,\bar{t}} = \sum_{v=0}^{n-1} C_{i_v j_v}^{0,\bar{t}},$$

$$where\ i_v = j_{v-1}, v = 1, 2, \ldots, n-1; j_{n-1} = i_v;$$

$$x_{i_v j_v}^{v+1,\bar{t}} = 1, v = 0, 1, \ldots, n-1;$$

The construction of the traveling salesman contours in directions $C^0 M(i_0, j_0^t)$, $t \in T = \{1, 2, \ldots, n-1\}$ begins at the nodes $v = 0, 1$ of the preliminary stage. From the operation of reducing the cost matrix $C^0 = \left[C_{ij}^0\right]_{n \times n} i \in T^0$, $j \in J^0$ characterizing the cost of the root node $i_0 \in I$, the elements restrictions of the reduced submatrix are carried out according to the scheme of operations.

The model is constructed from a sequence of recursive relationships that are inverse to each other at iterative steps. Recursive data and relationships are formed from elements of the cost matrix $C^0 = \left[C_{ij}^0\right]_{n \times n}$, $i \in I^0$, $j \in J^0$ and its reduced form $C^0R = \left[C_{ij}^{0r}\right]_{n\times n}$, $i \in I^0, j \in J^0$ with reduction coefficients $\alpha_i = minC_{ij}^0, i \in I^0, j \in J^0, \beta_j = min\left(C_{ij}^0 - \alpha_i\right), j \in J^0; i \in I^0$. The matrices C^0 and C^0R at the nodes of the tree contours are transformed into recursive submatrices with monotonically descending orders $(n - v)^2$:

$$C^v(i_{v-1}, j_{v-1}) = \left[C_{i_v j_v}^v\right]_{(n-v)(n-v)} i \in I^v, j \in J^v, v = 1, 2, \ldots, n-1;$$

$$i \in I^v, j \in J^v, v = 1, 2, \ldots, n-1; where I^v = I^{v-1} \backslash i_{v-1}; J^v = J^{v-1} \backslash j_{v-1};$$

In the recursive-decomposition model, the structure complexity of input and intermediate data is estimated by the following two criteria:

(1) length of the input and intermediate data code, measured by the orders of recursive submatrices $(n - 1)^2$, $v = 1, 2, \ldots, n - 1$;
(2) number of invalid items $C_{ij}^v = \infty\left(C_{ij}^{vr} = \infty\right) i \in I^v, j \in J^v$, $v = 0, 1, 2, \ldots, n - 1$;

Invalid elements implicitly reject subsets of traveling salesman contours with values greater than $f_{i_0} = f\left(x_{i_0}^n\right)$.

In addition, in the iterative algorithm work, these elements reduce the number of stationary admissible contours used to construct a minimal cost contour.

An iterative algorithm designed for a recursive-decomposition model works without internal loops. The length of the code, asymptotically, and the number of constructed contours estimate the computational complexity of the algorithm. The travelling salesman contour structure consists of a sequence of ($n - 1$) three-level nodes $i_v \in I^0$.

At the first two levels of each node $i_v \in I^v$, submatrices $C^v(i_{v-1}, j_{v-1})$ and $C^vR(i_{v-1}, j_{v-1})$ with restrictions are formed from the corresponding recursive submatrix $C^vR(i_{v-1}, j_{v-1})$.

At the third level of the node $i_v \in I^v$, the stationary admissible arc $(i_v, j_v) \in i_v \times J_{i_o}^v$ is determined. The arc changes the direction and length of the constructed contour $C^{v-1}M(i_0, j_0) = \left\{(i_k, j_k)_{k=0}^{v-1}\right\}$.$v = 1, 2, .., n - 1$. The traveling salesman contour $C^{n-1}M(i_0, j_0) = \{(i_v, j_v)_{v=0}^{n-1}\}$ starting from the node $i_0 \in I^0$ branch along the basic guide arcs $C^0M(i_0, j_0) = \left\{(i_0, j_0^t)\right\}$, $j_0 \in J_{i_0}^0 = J^0 \backslash i_0$ form a subset of stationary admissible contours. The subsets of the stationary admissible traveling salesman contours do not intersect in ($n - 1$) directions. The general objective function of these optimization subtasks is as follows:

$$f(x^{n-1}) = \sum_{v=0}^{n-1} C_{i_v j_v}^0 x_{i_v j_v}^{v+1} = \sum_{v=0}^{n-1} C_{i_v j_v}^0 => min,$$

where $x_{i_v j_v}^{v+1} = 1$, $v = 0, 1, .., n - 1$.

5 Conclusion

The complexity index of the algorithms is characterized by the growth order of time and memory capacity necessary to solve the problem with increasing input data. The time spent by the algorithm as a function of the size of the problem $F(n, f^n_{i_0,u=0})$ is called *time-cost complexity.*

The behavior of this complexity with an increase in the problem size is called *asymptotic time-cost complexity*. This determines the size of the problems that can be solved by an iterative algorithm, the problem of building the traveling salesman contours. The complexity index of the iterative algorithm is estimated by the growth order of the time necessary to solve the problem with increasing input data and the value of the initial cost $f^n_{i_0} = f(x^n_{i_0})$.

For a fixed $i_0 \in I^0$, the value of the cost $f^n_{i_0,u=0} = f\left(x^n_{i_0}\right) = \sum_{v=0}^{n-1} C^0_{i_v j_v}$, where $x^n = (x^{v+1}_{i_v j_v} = 1, v = 0, 1, \ldots, n-1)$ of the initial salesman contour $C^{n-1}_{u=0} M(i_0, j_0) = \left\{(i_v j_v)^{n-1}_{v=0}\right\}$, $j_0 \in J^0_{i_0}$ is the largest number in the implementing an iterative algorithm for constructing salesman contours. At the zero iteration step (v = 0), the value is used by the upper restriction of the elements of the cost matrix: $C^0 = \left[C^0_{ij}\right]_{n\times n}$, $C^0_{ij} < f_{i_0}$. Elements $C^0_{ij} \geq f^n_{i_0}$ equal to infinity: $C^0_{ij} = \infty$, $i \in I^0, j \in J^0$.

We introduce into the RD model the interval of admissible costs of stationary admissible contours $\Delta_0\left(f^n_{i_0}\right) = \left[\varphi^0, f_{i_0}\right]$, where $\varphi^0 = \alpha^0 + \beta^0$ is the sum of the reduction coefficients of the matrix $C^0 R$.

We created a new theorem, which state: In order for an MDG tree $C^{n-1}_{u-1} M(i_0 j_0)$ to be an optimal solution to the traveling salesman problem (TSP) in the recursive-optimization statement, it is necessary and sufficient that the subset $I_u(C^{0r}_{i_0})$ is empty, i.e. $I_u(C^{0r}_{i_0}) = \emptyset$.

An iterative algorithm with recursive data using the decomposition technique is effective. It is estimated by computational criteria for the complexity of processing the input size, the number of stationary admissible contours. In the root non-oriented tree describing the structure of the traveling salesman contours, the paths can go in the opposite direction with lengthening and shortening of the constructed contour segment.

References

1. Prim, R.C.: Shortest connection networks and some generalizations. BSTJ **36**, 1389–1401 (1957)
2. Gutin, G., Punnen, A.P.: The Traveling Salesman Problem and Its Variations, pp. 1–40. Kluwer, Dordrecht (2002)
3. Applegate, D.L., Bixby, R., Chvátal, V., Cook, W.J.: The Traveling Salesman Problem: A Computational Study. Princeton University Press (2007)
4. Weisstein, E.W.: Mathworld-A Wolfram Web Resources. Traveling Salesman Problem (2010)

5. Hurkens, C.A.J., Woeginger, G.J.: On the nearest neighbour rule for the traveling salesman problem. Oper. Res. Lett. **32**, 1–4 (2004)
6. Dantzig, G.B., Fulkerson, D.R., Johnson, S.M.: Solution of a large scale traveling salesman problem. Oper. Res. **2**, 393–410 (1954)
7. Cheriton, D., Tarjan, R.E.: Finding minimum spanning trees. J. SIAM Comp. **5**, 724–742 (1976)
8. Crowder, H., Padberg, M.W.: Solving large-scale symmetric travelling salesman problems to optimality. Manage. Sci. **26**, 495–509 (1980)
9. Goldberg, D.E., Lingle, R. Jr.: Alleles, loci, and the traveling salesman problem. In: Proceedings of International Conference on Genetic Algorithms and their Applications, pp. 154–159 (1985)
10. Arora, S.: Polynomial time approximation schemes for Euclidean traveling salesman and other geometric problems. J. ACM **45**, 753–782 (1998)
11. Held, M., Karp, R.M.: The traveling-salesman problem and minimum spanning trees. Oper. Res. **18**, 1138–1162 (1970)
12. Held, M., Karp, R.M.: The traveling-salesman problem and minimum spanning trees; part II. Math. Program. **1**, 6–25 (1971)
13. Papadimitriou, C.H., Yannakakis, M.: The complexity of restricted spanning tree problems. Automata, Languages, and Programming, pp. 460–470. Springer-Varlag, Berlin (1979)
14. Held, M., Karp, R.M.: A dynamic programming approach to sequencing problems. In: Proceedings of the 1961 16th ACM National Meeting, ACM '61, 71.201–71.204. New York, NY, USA, ACM (1961)
15. Bellman, R.: Dynamic programming treatment of the travelling salesman problem. J. ACM **9**(1), 61–63 (1962)
16. Bjorklund, A.: Determinant sums for undirected hamiltonicity. SIAM J. Comput. **43**(1), 280–299 (2010)
17. Woeginger, G.J.: Exact algorithms for NP-hard problems: a survey. Combinator. Opt. LNCS **2570**, 185–207 (2003)
18. Woeginger, G.J.: Open problems around exact algorithms. Discret. Appl. Math. **156**, 397–405 (2008)
19. Tucker, A.W., Miller, C.E., Zemlin, R.A.: Integer programming formulation and travelling salesman problems. J. ACM **7**, 326–329 (1960)
20. Little, J.D.C., Murty, K.G., Sweeny, D.W., Karel, C.: An algorithm for the traveling salesman problem. Oper. Res. **11**, 972–989 (1963)
21. Lin, S.: Computer solutions to the traveling salesman problem. BSTJ **44**(10), 2245–2269 (1965)
22. Cook, S.A.: The complexity of theorem proving procedures. In: Proceedings of 3rd ACM symposium on the theory of computing, ACM, pp. 151–158 (1971)
23. Karp, R.: Reducibility among combinatorial problems. In: Complexity of Computer Computations: Proceedings Symposium, pp. 85–103. Plenum Press, New York (1972)
24. Garey, M.R., Johnson, D.S.: Computers and Intractability: A Guide to the Theory of NP-completeness. W.H. Freeman and co., New York (1979)
25. Sahni, S., Gonzales, T.: P-complete approximation problems. J. ACM **23**, 555–565 (1976)
26. Papadimitriou, C.H.: The Euclidean traveling salesman problem is NP-complete. Theor. Comput. Sci. **4**, 237–244 (1977)
27. Kruskal, J.B.: On the shortest spanning subtree of a graph and the traveling salesman problem. Proc. Am. Math. Soc. **7**, 48–50 (1956)
28. Norman, R.Z., Rabin, M.O.: An algorithm for a minimum cover of a graph. Proc. Am. Math. Soc. **10**, 315–319 (1959)
29. Yao, A.C-C.: An O(|E|loglog|V|) Algorithm for Finding Minimum Spanning Trees. Inf. Proc. Lett. 4, 21–25 (1975)
30. Papadimitriou, C.H.: The complexity of the capacitated tree problem. Networks **8**, 217–230 (1978)
31. Papadimitriou, C.H., Vempala, S.: On the approximability of the traveling salesman problem. Combinatorica **26**, 101–120 (2006)
32. Reinelt, G.: The Traveling Salesman Computational Solutions for TSP Applications. Springer, Berlin (1994)

33. Dantzig, G.B.: On the significance of solving linear programming problems with some integer variables. Econometrica, 30–44 (1960)
34. Papadimitriou, C.H., Steiglitz, K.: Some examples of difficult traveling salesman problems. Oper. Res. **26**, 434–443 (1978)
35. Woeginger, G.J.: Exact algorithms for *NP-hard* problems. OPTIMA **68**, 2–8 (2002)
36. Valiant, L.G.: A Polynomial Reduction of Satisfiability to Hamiltonian Circuits that Preserves the Number of Solutions. University of Leeds, Manuscript (1974)
37. Khachian, L.G.: A polynomial algorithm for linear programming. Doklady Akad. Nauk USSR **244**(5), 1093–1096 (1979)
38. Papadimitriou, C.H., Steiglitz, K.: On the complexity of local search for the traveling salesman problem. SIAM J. Comput. **6**(1), 76–83 (1977)
39. Garey, M.R., Johnson, D.S.: Strong NP-completeness results: motivation, examples, and implications. J. ACM **25**, 499–508 (1978)
40. Edmonds, J.: Paths, trees, and flowers. Can. J. Math. **7**, 449–467 (1965)
41. Cobham, A.: The intrinsic computational difficulty of functions. In: Hillel, Y.B. (Ed.) Proceedings of International Congress for Logic Methodology and Philosophy of Science. Amsterdam, North Holland (1964)

The Change in the Main Characteristics of the company's Business Model on the Example of a Company Adal Catering

Galimkair Mutanov, Zhansaya Issayeva, and Askhat Serikbekuly

Abstract During the study, the main trends in the food technology market in the world and the impact of these trends on Kazakhstani companies in this industry were studied. Using as an example one of the Kazakhstani small business companies, the influence of world trends in food technology was studied, as well as other important factors that led to a change in the company's future business strategy, and analyzed more detailed changes in its important characteristics in the context of 9 business Canvas boxes. In addition, according to the results of the study, food delivery companies in the foodtech industry have a lot in common with FMCG distribution logistics companies, and in their activities they can rely on the principles and rules of this industry.

Keywords Foodtech · Food delivery · Business model canvas

1 Introduction

Recently, the trend of active growth in the world has such a direction as "foodtech". In a global sense, foodtech (food technology) is the integration of digital technologies throughout the food chain: from farms and food production to packaging, storage, preparation and disposal of food [1].

According to the report of Five Season Venture and dealroom.co [2] over the past 4 years, the volume of venture investments in foodtech in the world has grown almost 5 times: from € 3.08 billion in 2014 to € 14.3 billion in 2018 [2].

G. Mutanov · Z. Issayeva · A. Serikbekuly
Al-Farabi Kazakh National University, Almaty, Kazakhstan

A. Serikbekuly (✉)
UP Consulting, Almaty, Kazakhstan
e-mail: sa@analytic.kz

G. Mutanov
Institute of Information and Computational Technologies CS MES RK, Almaty, Kazakhstan

G. Mutanov and A. Serikbekuly (eds.), *Digital Transformation in Sustainable Value Chains and Innovative Infrastructures*, Studies in Systems, Decision and Control 443,
https://doi.org/10.1007/978-3-031-07067-9_13

Table 1 Data on the number of food delivery service users

Service name	Number of android users
Uber Eats	>100 000 000
Glovo	>10 000 000
DoorDash	>10 000 000
Deliveroo	>10 000 000
Foodpanda	>10 000 000
Takeaway.com	>10 000 000
DeliveryClub	>10 000 000
Wolt	>1 000 000
Zakazaka	>1 000 000
Instacart	>1 000 000
Yandex.Eda	>1 000 000
SberMarket	>500 000
Chocofood	>100 000

Today, the development of foodtech is expressed in the use of new methods and technologies for the production of products (growing vegetables and fruits; production of ready-made dishes); ways of packaging and storage of products, allowing to increase the shelf life of products and preserve their useful properties; methods and technologies used in waste management and so on.

But often the term foodtech refers to online food delivery services from shops and restaurants. This is due to the fact that it is this area that enjoys the greatest popularity and active growth rates today. About 60% of investments in world food technology are in food delivery via the Internet from shops and restaurants. American DoorDash, Instacart, Uber Eats, European Deliveroo, Delivery Hero, Takeaway.com are developing in this market [2].

These trends also affect the EurAsEC countries; these areas are particularly actively developing in Russia and Kazakhstan.

In Russia, companies such as Yandex (Yandex.Eda service) and Mail.Ru Group (DeliveryClub, Zakazaka and SberMarket services) mainly compete in the field of online delivery [2].

Kazakhstan consumers are also actively using food delivery services. Today, the largest players in this direction in the Kazakhstan market are such services as Glovo, Wolt, Yandex.Eda and domestic Chocofood.

The study examined the data on the number of application downloads[1] for each of the food delivery services mentioned in Table 1.

Such market trends are encouraging Kazakhstani producers (restaurants, cafes, canteens, fast food restaurants) to rebuild their business processes. In particular, manufacturing companies, in order to ensure the closure of the entire cycle of the

[1] Data on the number of downloads was taken only for users of the Android operating system from the official Google Play Android application store.

customer service process, are compelled, in addition to production, to carry out the delivery of their products to the final consumer. Such changes lead to the need for global changes in all business processes of the company. The consequence of such changes can be a fundamental change in the basic characteristics of the company according to the Business Model Canvas [3].

2 Literature Review

2.1 Business Model Canvas

Business Model Canvas (BMC) is a strategic management and lean startup template for developing new or documenting existing business models [4, 5]. It is a visual chart with elements describing a firm's or product's value proposition, infrastructure, customers, and finances [3].

BMC has nine boxes, a more detailed description of which is presented below [3, 6]:

1. Key Activities: The most important activities of the company, providing the value proposition.
2. Key Resources: Company assets that are necessary to support the business and ensure its functioning. These resources are involved in creating value for the customer.
3. Key Partners: Partners help the company to develop successfully, as well as optimize operations and reduce the risks of the business model.
4. Value Propositions: Products and services offered by the company to meet the needs of its customers, which distinguish it from its competitors. These are the values that customers are willing to pay for.
5. Customer Segments: Clients can be divided into groups based on their various needs and preferences, in order to ensure the proper implementation of the corporate strategy.
6. Channels: The channels through which the company's value proposition will be distributed to its target customers in the most effective ways.
7. Customer Relationships: This box defines the main areas of interaction with customers: how the company plans to attract new customers, how to retain existing customers, and how to earn income from its current customers.
8. Cost Structure: all available cost items that the company has in carrying out its activities.
9. Revenue Streams: The main sources of revenue for each of the customer segments.

The Business Model Canvas is the so-called a roadmap for all businesses. Each of the existing and functioning companies can use Business Model Canvas to study its product or service starting from any of 9 boxes [7].

In his work, Stenn [7] talks about such a concept as the "Sustainable Development Lens". The Sustainability Lens is part of a larger body of thought called Design Thinking. Design Thinking is an interactive, nonlinear, visual way of looking at how things connect [7].

The BMC is also a design thinking tool developed to create more access to business thinking, planning, strategizing and problem solving. BMC and Sustainability Lens share one common goal—to provide opportunities to increase business resilience [7].

The Canvas business model is a business design tool that is gaining in popularity, especially for new market players. BMC's global goal is to strike a balance between its elements—effective work with suppliers, customer relations and the ability to rationally manage internal resources [8].

Value proposition is the central box of BMC. It describes the products and services provided by the company to meet the needs of its customers (from one or more different segments). By developing various channels to promote and distribute its products or services, the company builds communication with its customers, providing the desired value [9].

The "Customer Relationships" block describes specific areas of relationships with various customer segments, as well as ways to achieve mutually beneficial relationships, including their acquisition and further retention. Effective interaction between these blocks provides revenue streams, which include cash received from customers [9].

Another part of the Business Model Canvas, including key partners, key resources and key activities, is the basis for creating the value of a product or service and constitutes a cost structure [9].

2.2 *Food Delivery*

There is a lot of scientific work on the active development of such areas as online food delivery. For example, Lau Tech Chai and David Ng Ching Yat [10] write that the emergence and development of online food delivery services is due to the changing nature of the needs and preferences of urban consumers. The most common and significant reason for the growth in popularity of such services is their ability to quickly provide customers with lunch during the working day or dinner after it [10].

In turn, food delivery services also have a certain impact on consumer behavior and the use of these services is becoming the norm [10].

Tamaraiselvan N., Jayadevan G.R. and Chandrasekar K.S. say that the business of delivering food from restaurants is undergoing enormous changes [11]. Not only services for the delivery of prepared food from restaurants and cafes, but also online platforms for the delivery of food and drinks are beginning to actively develop and capture the market [10, 11].

According to McKinsey, this market has already matured in many countries by 2016, and its growth rate was estimated at 3.5 percent over five years [12].

Triyana C. and Syaroni D. also said that one of the fastest growing businesses currently is food delivery via the Internet [13].

Panse C., Rastogi Sh., Sharma A., Dorji N. in their article describe the impact of the use of online food delivery services on restaurants, and also talk about such a concept as cloud kitchens [14].

Among the positive effects of using online food delivery services for restaurants:

- When customers are satisfied, this application can help restaurants develop customer loyalty, increase profitability and enter new market segments [14];
- If there is a lot of traffic in restaurants, this application helps them get consumers, and not lose because of the long waiting time [14];
- If restaurants, on the contrary, have low traffic, this application may open up opportunities to find new customers and businesses, since ordering food online is much more convenient and cheaper than having dinner outside the house;
- Accommodation in the application is beneficial for restaurants, since the application promotes the restaurant on its own behalf, which leads to an increase in the customer base and lower own costs for promotion.

The negative effect is that the fine dines outlook is mainly on ambience and taste—ordering food through a mobile application can only be based on price and convenience.

Cloud kitchen—it is also called a delivery-only restaurant. They are small eateries where there is a common kitchen, allowing delivery. This helps brands get closer to customers and lower the cost of cooking.

The growing popularity of cloud kitchens was mainly due to factors such as the complexity of creating and maintaining the restaurant premises and the emergence of technologies that allow home delivery of products. Cloud kitchens eliminate the need to serve visitors on site and allow you to settle in less expensive places [14].

3 Methods

The object of the study is a small sized company in food delivery—"Adal Catering".

To obtain data on business processes and the main characteristics of the company at the initial stage of the study, as well as after changing the business model, a detailed analysis of the company was carried out using such research methods as observation and interviews.

During the observation by the researcher empirically, being in the canteen at the initial stage, and on the territory of the production workshop and administration office, basic information was obtained on how the company's activities were organized.

In the course of the interview with the owner of the company and its employees, additional more detailed information about the company's operations was obtained.

The main tool used in the study is the Business Model Canvas [3].

As the results of the study, a description of the Business Model Canvas boxes at the time the study began and the final version, after changing the direction of the company.

4 Results and Discussion

At the time of the start of the study, Adal Catering was a canteen whose main activity was the production and sale of its own products offline. Under the influence of changes in the food industry not only in the world, but also in Kazakhstan, the company was forced to revise its business strategy and resort to fast reengineering—this phenomenon has been studied in detail and described by Mutanov and Serikbekuly [15].

The company's characteristics for the Canvas business model at the time the research began were as follows:

1. Key activities:
   - production of ready meals;
   - customer service.
2. Key resources:
   - premises (real estate);
   - kitchen equipment (stoves, refrigerators);
   - raw materials for food production—products;
   - cafe staff (cooks, waiters, administrator);
   - production accounting system (IIKO).
3. Key partners:
   - –
4. Value Propositions:
   - affordable price.
5. Customer Segments:
   - employees of the nearest offices.
6. Channels:
   - cafe (offline).
7. Customer Relationships:
   - service at the time of visiting the cafe.
8. Cost Structure:
   - rental of premises;

- utilities;
- purchase of raw materials (products);
- the salary of the cafe staff;
- depreciation of equipment.

9. Revenue Streams:

   - margin for the sale of ready meals;
   - service.

Figure 1 shows a graphical representation of the Adal Catering Business Model Canvas at the initial stage of the study.

The key factors that led to the need for reengineering and changes in the company's business model were:

- Changing the behavior and preferences of consumers—it is important for the consumer to be able to have lunch / dinner quickly and conveniently—in most cases, without leaving home. People want fresh and wholesome foods without the effort.

<table>
<tr><td rowspan="2">Key Partners:</td><td>Key Activities:

- production of ready meals
- customer service</td><td rowspan="2">Value Propositions:

- affordable price</td><td>Customer Relationships:

- service at the time of visiting the canteen</td><td rowspan="2">Customer Segments:

- employees of the nearest offices</td></tr>
<tr><td>Key Resources:

- premises (real estate)
- kitchen equipment (stoves, refrigerators)
- raw materials for food production - products
- cafe staff (cooks, waiters, administrator)
- production accounting system (IIKO)</td><td>Channels:

- cafe (offline)</td></tr>
<tr><td colspan="3">Cost Structure:

- rental of premises
- utilities
- purchase of raw materials (products)
- the salary of the cafe staff
- depreciation of equipment</td><td colspan="2">Revenue Streams:

- margin for the sale of ready meals
- service</td></tr>
</table>

Fig. 1 Business model canvas at the initial stage of research

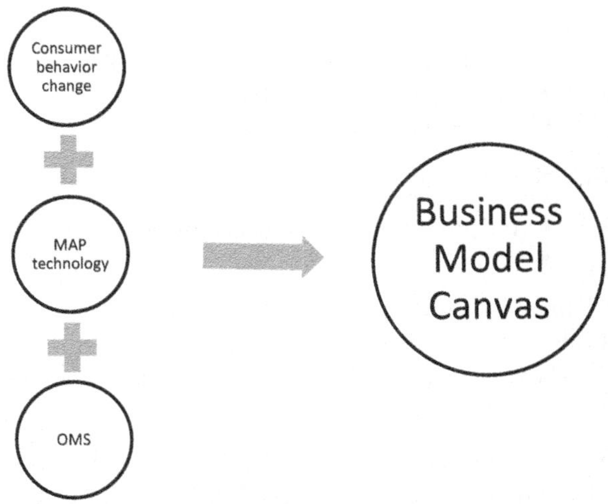

Fig. 2 Factors affecting a company's business model (BMC)

- Application of MAP (Modified Atmosphere Packaging) technology. The technology is based on the packaging of products in a specialized gas environment, which can significantly increase the shelf life of the finished product.
- Using an IT service in work that allows you to efficiently organize logistics, including receiving orders and delivering finished products to customers—OMS (Order Management System).

As an OMS, company used IT service (EZAPP) intended for distribution companies in the FMCG sector, which was adapted for the delivery of ready-made meals. Adapting the service to the food delivery activities was carried out promptly and successfully, as this business model has much in common with the model of FMCG delivery companies.

The influence of the described factors on the company's business model is presented schematically in Fig. 2.

The characteristics of the company after changing the direction of activity in the context of Business Model Canvas boxes today are as follows:

1. Key activities:
    - production of ready meals;
    - receiving orders from customers;
    - delivery of orders.
2. Key resources:
    - production room;
    - kitchen equipment (stoves, refrigerators);
    - MAP technology packing conveyor;
    - raw materials for food production—products;
    - order management system (OMS—EZAPP).

3. Key partners:
    - administration of business centers.
4. Value Propositions:
    - affordable price;
    - free city delivery;
    - availability at the nearest retail outlets;
    - long shelf life of the product.
5. Customer Segments:
    - employees of offices and business centers of the city;
    - housewives;
    - seniors;
    - students;
    - retail outlets.
6. Channels:
    - retail outlets;
    - own delivery.
7. Customer Relationships:
    - taking orders by phone;
    - receiving orders online through the order management system;
    - interaction at the time of the visit to the outlet.
8. Cost Structure:
    - rental of industrial premises;
    - office rental;
    - utilities;
    - purchase of raw materials (products);
    - production staff salary;
    - salary of sales and logistics personnel;
    - monthly payment order management system;
    - depreciation of equipment.
    - logistics costs;
9. Revenue Streams:
    - margin for the sale of ready meals.

A graphical representation of the Adal Catering Canvas business model after changing the business model is shown in Fig. 3.

As a result of the changes that the company has undergone, there have been changes in all 9 BMC boxes. The main changes in each of the boxes are presented in comparative Table 2.

Considering the changes that have occurred in more detail, they are:

<table>
<tr>
<td rowspan="2">Key Partners:

- administration of business centers</td>
<td>Key Activities:

- production of ready meals
- receiving orders from customers
- delivery of orders.</td>
<td rowspan="2">Value Propositions:

- affordable price
- free city delivery
- availability at the nearest retail outlets
- long shelf life of the product</td>
<td>Customer Relationships:

- taking orders by phone
- receiving orders online through the order management system
- interaction at the time of the visit to the outlet</td>
<td rowspan="2">Customer Segments:

- employees of offices and business centers of the city
- housewives
- seniors
- students
- retail outlets</td>
</tr>
<tr>
<td>Key Resources:

- production room
- kitchen equipment (stoves, refrigerators)
- MAP technology packing conveyor
- raw materials for food production - products
- order management system (OMS - EZAPP)</td>
<td>Channels:

- retail outlets
- own delivery</td>
</tr>
<tr>
<td colspan="3">Cost Structure:

- rental of industrial premises
- office rental
- utilities
- purchase of raw materials (products)
- production staff salary
- salary of sales and logistics personnel
- monthly payment order management system
- logistics costs
- depreciation of equipment</td>
<td colspan="2">Revenue Streams:

- margin for the sale of ready meals</td>
</tr>
</table>

Fig. 3 Business model canvas after a change in direction of activity

1. The list of client segments has become much wider, due to the fact that the geographical confinement is now determined by the delivery zone (city boundaries), and not by the actual location of the canteen (micro-district border).
2. The list of value offers that the company can offer its customers has significantly expanded—products have become more accessible to customers due to delivery and product availability in nearby stores. One of the most important advantages

Table 2 Main changes in BMC boxes after changing the direction of activity

No.	BMC block name	Before changes	After changes
1	Customer Segments	– employees of the nearest offices	– employees of offices and business centers of the city – housewives – seniors – students – retail outlets
2	Value Propositions	– affordable price	– affordable price – free city delivery – availability at the nearest retail outlets – long shelf life of the product
3	Channels	– canteen (offline)	– retail outlets – own delivery
4	Customer Relationships	– service at the time of visiting the canteen	– taking orders by phone – receiving orders online through the order management system – interaction at the time of the visit to the outlet
5	Revenue Streams	– sale of ready meals – service	– sale of ready meals
6	Key Resources	– premises (real estate) – kitchen equipment (stoves, refrigerators) – raw materials for food production (products) – cafe staff (cooks, waiters, administrator) – production accounting system (IIKO)	– production room – kitchen equipment (stoves, refrigerators) – raw materials for food production (products) – MAP technology packing conveyor – order management system (OMS—EZAPP)
7	Key Activities	– production of ready meals – customer service	– production of ready meals – receiving orders from customers – delivery of orders
8	Key Partners	(none)	– administration of business centers
9	Cost Structure	– rental of canteen premises – utilities – purchase of raw materials (products) – depreciation of equipment – the salary of the cafe staff	– rental of cooking premises – utilities – purchase of raw materials (products) – depreciation of equipment – production staff salary – salary of sales and logistics personnel – monthly payment order management system – logistics costs – office rental

was the increase in the shelf life of the finished product through the use of Modified Atmosphere Packaging technology.

3. Instead of one channel—an offline cafe, there were many—hundreds of retail outlets in the city where the company's products are available for sale, and their own delivery, which allows you to distribute ready-made meals to many business centers and private houses and apartments.
4. Previously, the only channel for interacting with customers was the customer service process in the dining room, but now the customer can order and receive goods in any way convenient for him.
5. The company began to receive income only from sales of finished products, but due to changes in the cost structure, the price of the final product remained the same.
6. The set of key resources has changed—the company abandoned such a resource as canteen staff and canteen premise, which excluded the costs of its maintenance from the general cost structure. Important resources such as MAP technology packing conveyor and order management system (OMS) were added, which were one of the main prerequisites for revising the company's strategy and changing its direction of activity.
7. Key actions such as receiving and delivering orders were added, which opened up new possibilities for expanding the list of value offers, as well as expanding sales markets.
8. With the new organization of its activities, the company has found partners represented by the administrators of business centers who are actively involved in promoting the Adal Catering products among its employees.
9. The cost structure has changed—the cost of maintaining the staff of canteen premise and canteen employees has been replaced by the cost of delivery, maintaining the staff of the sales and logistics departments and the order management system.

One of the important conclusions of the study is the fact that following market trends and undergoing factors that have a significant impact on the company (such as changing consumer behavior, the use of new technologies in the production and organization of business processes), and changing fundamentally As a result of its business model and activity direction, the food tech company ultimately has one of its key goals as an efficient organization of distribution logistics.

Distribution logistics, in turn, sets as its main goal the availability of the necessary goods (finished product, in this case), available to the consumer at the right time and in the right amount [15].

Thus, the modern foodtech industry is becoming similar to the FMCG logistics industry, which in the future will make it possible to apply the methods and rules applicable to FMCG logistics to companies that organize the delivery of ready meals to their customers.

5 Conclusion

Drived by market trends edgy and global changes in the foodtech market, small companies are often forced to resort to significant changes in their activities and revise their strategy in order to continue their existence and further successful development.

In addition to the influence of market trends and changing consumer preferences, the acquisition and use of MAP technology equipment, which allows for a longer shelf life of products through specialized packaging technology, played a decisive role for the company in changing its profile.

Also, one of the factors was the use of the company's order management system, which allowed finding new ways to interact with existing and new customers and more efficiently organizing the order processing process.

After analyzing all the changes that have occurred in the company after changing the direction of activity through the prism of Business Model Canvas, there is an opportunity to see how each of the strategic blocks for the company has changed and to understand what new opportunities the company may have in the future, due to which they may appear and what the company will bring in the end the result.

In addition, it was revealed that the companies in the finished food delivery industry share the same characteristics as the FMCG distribution logistics companies.

In a rapidly changing market, business models can change dramatically, which requires constant monitoring of changes in existing business models, the reasons and factors affecting these changes, as well as the emergence of new business models for food delivery.

The reasons for further changes in the footwear industry may be:

- Investments of venture funds in the food industry;
- Changes in consumer behavior and preferences;
- Change in existing and the emergence of new technologies in the production, processing and delivery of products.

All these factors, their influence on each other and on the food tech market, need further monitoring and more detailed study in future studies.

Acknowledgments This research has been funded by the Science Committee of the Ministry of Education and Science of the Republic of Kazakhstan (Grant No. AP08052658).

References

1. Vc.ru Nine promising food tech areas from an investor perspective. https://www.vc.ru/food/60964-devyat-perspektivnyh-fudteh-napravleniy-s-tochki-zreniya-investorov. Accessed 05 Feb 2020
2. Dealroom The State of European Food Tech 2018. https://www.blog.dealroom.co/the-state-of-european-food-tech-2018/. Accessed 05 Feb 2020

3. Osterwalder, A., Pigneur, Y.: Business Model Generation: A Handbook for Visionaries, Game Changers, and Challengers. Strategyzer series. John Wiley & Sons., Hoboken, NJ (2010)
4. APB Barquet VP Cunha MG Oliveira H Rozenfeld 2011 Business model elements for product-service system J Hesselbach C Herrmann Eds Functional Thinking for Value Creation Springer Berlin, Heidelberg 332 337
5. M Reuver H Bouwman T Haaker 2013 Business model roadmapping: a practical approach to come from an existing to a desired business model Int. J. Innov. Manage. 17 1 18
6. Osterwalder, A.: The business model ontology: a proposition in a design science approach. Doctoral Thesis, University of Lausanne, Lausanne, Switzerland (2004)
7. T Stenn 2017 Social Entrepreneurship as Sustainable Development 1 Palgrave Macmillan Cham
8. Prasetyawan, Y., Maulida, N., Lutvitasari, M.R.: The integration between Business Model Canvas and Manufacturing System Design. In: International Conference on Industrial and System Engineering. p. 337. IOP Publishing (2018)
9. Romero, M.C., Villalobos, J., Sanchez, M.: Simulating the business model canvas using system dynamics. In: 10th Computing Colombian Conference, pp. 527–534. Bogota (2015)
10. Lau, T.C., David, N.G.: Online food delivery services: making food delivery the new normal. JMAP **1**(1), 62–77 (2019)
11. N Thamaraiselvan GR Jayadevan KS Chandrasekar 2019 Digital food delivery apps revolutionizing food products marketing in India Int. J. Recent Technol. Eng. 8 256 662 665
12. Mckinsey The changing market for food delivery. https://www.mckinsey.com/industries/high-tech/our-insights/the-changing-market-for-food-delivery. Accessed 08 Feb 2020
13. Triyana, C., Syaroni, D.: Consumer perception about paid cutlery programs at online food delivery company. In: ICOBEST-EBM 2019, pp. 89–93. Atlantis Press SARL (2020)
14. C Panse SH Rastogi A Sharma N Dorji 2019 Understanding consumer behavior towards utilization of online food delivery platforms J. Theor. Appl. Inf. Technol. 97 16 4353 4365
15. Mutanov, G.M., Serikbekuly, A.: Fast reengineering for distribution logistic processes in small companies in Kazakhstan based on graphical and system dynamic model. In: Proceedings of the 34th International Business Information Management Association Conference (IBIMA) "Vision 2025: Education Excellence and Management of Innovations through Sustainable Economic Competitive Advantage", Madrid, Spain, pp. 11687–11697 (2019)

Features and Problems of Marketing Services in the Digital Age

Zaresh Akhmetova, Moldir Belgozhakyzy, and Ella Pak

Abstract Service marketing arose in response to a shortage of product marketing. Although earlier traces exist, they became widespread in the mid- 70s. the evolution of the service sector can be divided into phases in which critical incidents can be identified that have led the service sector in new directions. The Central place in the discipline is occupied by a service meeting. Service marketing plays a role for the growth of any organization. The importance of marketing services is growing day by day. Along with the need for better service quality increases with customer expectation. This area can no longer be compromised. This article is an attempt to understand the importance of service marketing and quality of service. Various service gaps were identified: reliability, responsiveness, competence, access, courtesy, communication, trust, security, understanding, and material capabilities. The research discussed here consists of ways to understand the customer's experience when meeting the service and the consequences of that meeting. We summarize the various methods or approaches that have been and are used to understand the meeting with the service.

Keywords Marketing of services · Customer needs · Marketing research · Relationship with the consumer

1 Introduction

Currently, the world economy is increasingly characterized as a service economy. This is primarily due to the growing importance and share of the service sector in the economy of most developed and developing countries. In fact, the growth of the service sector has long been considered an indicator of the country's economic

Z. Akhmetova (✉) · E. Pak
Al-Farabi Kazakh National University, al-Farabi Ave. 71, 050040 Almaty, Republic of Kazakhstan
e-mail: Axmetova.Zauresh@kaznu.edu.kz

M. Belgozhakyzy
University of International Business, Abaya Ave. 8a, 050010 Almaty, Republic of Kazakhstan

G. Mutanov and A. Serikbekuly (eds.), *Digital Transformation in Sustainable Value Chains and Innovative Infrastructures*, Studies in Systems, Decision and Control 443,
https://doi.org/10.1007/978-3-031-07067-9_14

progress. Economic history tells us that all developing countries have consistently experienced the transition from agriculture to industry, and then to the service sector as the main component of the economy [1]. The service sector includes industries such as banking, insurance, communications, consulting, non-profit activities, travel and transportation, and all other businesses that do not produce tangible goods.

Service marketing management became a separate area of research only in the early 1980s, when marketers finally realized the special characteristics of service marketing management. Simply put, service marketing management deals with actions and processes that allow a service provider to provide services to end users.

The service sector is likely to be the most important sector in any developed economy. In developed countries, the service sector currently generates more than 70% of the country's GDP [2]. The importance of the service economy continues to grow; the largest growth in the number of firms is in the service sector, which in turn means that the number of employees in this sector is also increasing. New, more service-oriented business models are being introduced, which are seen in manufacturing companies that are turning into service providers. For example, IBM has completely changed its business so that it no longer produces goods, but only produces and provides services.

This shift has also led to a change in the definition of goods and services themselves. The product is no longer considered separate from the services. Rather, services are now increasingly an integral part of a product, and this interconnectedness of goods and services is represented in a continuum of goods and services.

We also see a growing economy of collaboration, and companies like Airbnb, Uber, and Craigslist are changing the rules of competition in the service industry [3]. In addition, business models are changing in retail, where companies like AliExpress and Amazon are forcing changes in competition. To survive, companies must be able to provide a good service with a perception of high quality adapted to the needs of customers. But this is just a prerequisite; to be competitive, they have to go even further. Companies must understand and improve customer interaction and experience in all areas of any service. Therefore, it is impossible to ignore the fact that the service sector is becoming increasingly important and will continue to be relevant. To be relevant, researchers and organizations must develop new and better tools built to better understand the client's point of view. Companies need to balance how much their customers want the organization to know about them with what they need to learn. It is a fact that organizations can acquire very deep knowledge about their customers with the help of still developing technologies.

2 Literature Review

The American marketing Association defines services as activities, benefits, or satisfaction that are offered for sale or provided with the sale of goods to the buyer, i.e., pre-sales and after-sales services. Berry States that "although a product is an object, device, or physical thing, a service is a matter, result, or effort".

Service researchers usually rely on Shostak's definition [4], meeting with services as "the period of time during which the consumer directly interacts with the service". It is during these meetings that customers form their perception of the quality of service. Since customers encounter multiple contacts not only directly, but also indirectly through advertising or social networks, different contacts can lead to the formation of loyalty and relationships. The main focus of a meeting with a service is usually on the main meeting, but what happens before and after the main meeting also affects how customers perceive the main meeting with the service and the overall relationship with the organization [5].

A very important and highly cited concept was The Parasuraman, Zeithaml and Berry gap model [6]. The gap model identifies five gaps—"these gaps may be the main obstacles in trying to provide a service that consumers would perceive as high-quality" [6]. The natural conclusion from these concepts was that the quality of services is seen as something different from the quality of products and should be measured and managed differently. For example, quality attributes can be classified as search, experience, and trust attributes [7]. For products, search attributes (attributes that can be defined before purchase) are usually more important, while for services, experience, and trust, attributes are more important.

Concepts, frameworks, and strategies for marketing services were developed as a result of the interaction of many industries, organizations, and individuals who realized the increasingly important role that services play in today's global economy. Initially, the development of service marketing was focused on service industries. However, the industrial and technology industries have recognized services as a prerequisite for complementing their products in order to successfully compete in the market. Thus, it can be argued that in most industries, the provision of services is no longer an option, but a necessity.

In their study, Ziethaml saw differences in service marketing. They noted that marketing firms to institutional clients differs from marketing firms to end users in several important ways. For example, advertising seems to be a less important part of the marketing program of institutional firms. They seem to be more interested in contact marketing with a strong focus on customer follow-up. This is due to the fact that there are fewer clients in the institutional market who also tend to spend more [8]. In early works on services, scientists distinguished services from goods, noting that they are immaterial, perishable, changeable, and that the producer and consumer are inseparable. It has recently been suggested that these distinctive characteristics should not be considered unique to services, but that they also relate to goods, that "all products are services", and that "economic exchange is mainly related to the provision of services" [9].

3 Methodology

The methodological basis of the scientific article includes:

1 the philosophical method on which analytical conclusions and theories about the development of the concept of "service marketing" are based;
2 scientific method: the authors analyzed the world experience of introduction of the service marketing;
3 classification and comparative method: the features and main problems of service marketing were considered.

4 Results

The service concept is the entire package that your customer buys. For example, a consumer who orders a meeting at a law firm can expect the meeting to take place in a professional environment. The person opening the Bank account is waiting for ATMs and online access to their account. Product marketers define a product concept as a marketing Suite or 4P Product, Price, Promotion, and Placement. They try to create the perfect marketing package before bringing the product to market, so customers are more likely to buy it directly from the shelf. But in service marketing management, the real game begins only after the consumer meets the service provider. This makes the work of service marketers much broader and extends the concept of service beyond 4P.

Service marketing management includes 3 additional Components: People, Process, and Physical data.

- People include employees who provide the service. Since no company has ever made happy customers unhappy or incapable employees, it will be necessary to train their employees well and ensure their job satisfaction, including financially. Employees will have to work with consumers of services, which you should also treat as your own people-remember, they are involved in the process of providing services.
- A process is a service delivery process that we are going to create in the next step.
- Physical evidence is the environment in which the service is provided. It includes the physical object where the service is produced and sold (Fig. 1).

Features of services:

1. Immateriality:

   The physical product is visible and concrete. Non-material services. The service cannot be touched or viewed, so it is difficult for customers to tell in advance what they will receive. For example, banks promote the sale of credit cards by emphasizing the convenience and benefits associated with owning a credit card.
2. Inseparability:

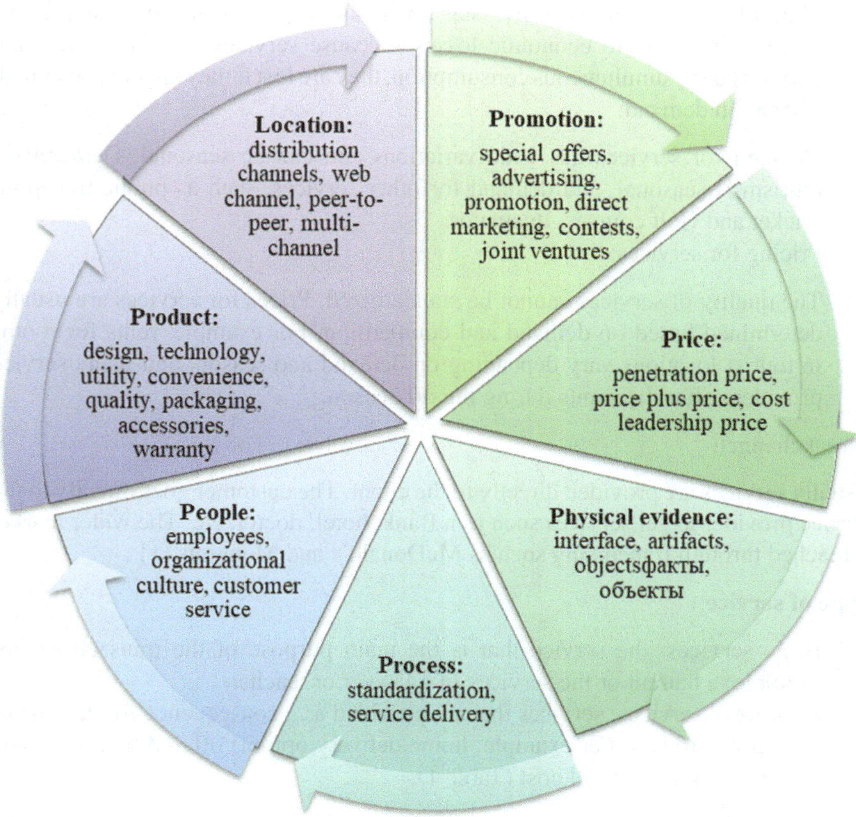

Fig. 1 Service marketing management components

Personal services cannot be separated from the individual. Services are created and consumed simultaneously. The service is created at the same time as the client receives it; for example, during an online search or legal consultation. Dentists, musicians, dancers, etc. Create and offer services simultaneously [10].

3. Heterogeneity (or variability):

   Services attract people, and people are all different. There is a high probability that different people (or even the same person at different times) will respond to the same query somewhat differently. It is important to minimize differences in performance (through training, standard setting, and quality assurance). The quality of the services offered by the companies can never be standardized.

4. Perishable:

   Services have a high degree of perishable products. Unused capacity cannot be saved for future use. If services are not used today, they are lost forever. For example, spare seats on an airplane cannot be transferred to the next flight.

Similarly, empty rooms in five-star hotels and unused loans are examples of services that lead to economic losses. Because services are actions that are performed for simultaneous consumption, they are lost if they are not consumed.

5. Change in demand:

   Demand for services has wide variations and can be seasonal. Demand for tourism is seasonal, and demand for other services, such as public transport, cricket and Golf courses, fluctuates.

6. Pricing for services:

   The quality of services cannot be standardized. Prices for services are usually determined based on demand and competition. For example, rents for rooms in tourist locations vary depending on demand and season, and many service providers offer discounts during the off-season.

Direct channel:

Usually services are provided directly to the client. The customer goes directly to the service provider to get services such as a Bank, hotel, doctor, etc. The wider market is reached through franchising such as McDonald's and Monginis [1].

Type of service:

1. Basic services: the service that is the main purpose of the transaction. For example: a haircut or the services of a lawyer or teacher.
2. Additional services: services that are provided as a consequence for the sale of a tangible product. For example: home delivery options offered by restaurants above the minimum bill cost (Table 1).

Problems in marketing services:

1. The service cannot be demonstrated.
2. Sales, production, and consumption of services occur simultaneously.
3. The service cannot be saved. This cannot be done while waiting for demand.
4. Services cannot be protected by patents.
5. Services cannot be separated from the service provider.
6. Services are not standardized and inconsistent.

Table 1 The main differences between physical goods and services

Services	Goods
Process or activity	Physical goods
intangible	tangible
Production, distribution, and consumption are simultaneous processes	Production and distribution is a separation from their consumption
Cannot be saved	Can be stored
Transfer of ownership is not possible	Possible transfer of ownership

7. The service providers are appointing franchisees may face challenges in the quality of services.
8. The customer's perception of service quality is directly related to the morale, motivation, and skills of the staff at the forefront of any service organization [1].

The difficulties encountered in providing technical assistance are as follows: Technical assistance for most companies is part of the product itself. Therefore, they must have a highly qualified and experienced technical team. An industrial marketing firm must have an open mind and a positive attitude to serving its customers.

Many marketers start with the position that customers are unhappy and always complain. Because of this relationship, the focus is not on services, and marketers believe that once the product is sold and the sale is closed, the work is completed. Since individual marketing is more like relationship marketing, it is necessary to maintain a good relationship with customers, because by now we are well aware that it is much cheaper to maintain the current set of customers than to attract new customers.

Due to the characteristics of the services, providing technical assistance can be difficult. Below are the characteristics of the services, as well as clarification of how they create difficulties in providing technical assistance [11].

1. Immateriality:

   The service is intangible, meaning you can't see, feel, or try it before you buy it. As a marketing specialist, you need to be aware of the importance that the customer places on a part of the service and project it in ads and promotions to attract the attention of customers and satisfy them. This immateriality creates certain difficulties. Since it cannot be seen or felt before it is tested, it is difficult to quantify and get a price. Hence the attempt to use material variables and quantitative evaluation so that any technical assistance can also bring profit to industrial marketing along with customer satisfaction [8].
2. Inseparability:

   Physical goods are produced on site, pass through the channel, and finally reach the customer. But on the other hand, if a person provides a service, the provider is part of that service. Therefore, the supplier-client relationship is important. The supplier must also show that the client can be processed [10]. In a product that is highly technical equipment, the manufacturer must provide technical training on how the machines should work. This is another type of technical assistance. If the trainer sent by the manufacturer does not have this option, then due to the "inseparable" nature of the service, this will affect the service and then the product itself. The manufacturer will finally be affected.
3. Variability:

   This is a small change from the previous characteristic. Because the service depends on who provides it, there is a high degree of variability, even if the same type of service is provided. This variability is a reflection of quality. Therefore, even services must follow certain quality control measures. The company can standardize the service by setting up specific systems and procedures so that all customers receive the same level of assistance and are satisfied [12].

4. Fluctuations in Demand:

   When demand is stable, services can be easily completed in advance. When demand fluctuates, an industrial firm creates problems. The firm of tax consultants will have more work at the end of the financial year and will therefore need additional experienced staff for this particular period. On the other hand, the firm may not have additional employees throughout the year because they will not work enough due to less demand. Thus, demand needs strategic forecasting and the development of an alternative plan. For example, part-time employees or consultants may be hired by the firm for short periods of time [13].

Marketing solutions to problems:

(1) in order to sell services, companies must reduce anxiety and build the trust of their customers. A great place to start is branding. A strong brand identity that mimics the company's defining qualities can give a sense of touch [14]. It is also important to ensure that branding and messaging remain consistent between channels and the media, as this shows the company's commitment to detail and quality.

    The concept of "show, don't tell" is an integral part of effective service marketing. Customer reviews, case studies, and reviews are all effective ways to communicate your value to potential customers and establish mutual understanding. Service businesses can also build trust by offering insight and insight into their process and people. Blogs, official documents, and articles are a great way to showcase your employees' knowledge and experience and start building relationships with customers. However, materiality must go beyond branding and advertising materials.

    Marketing represents the customer's voice in the organization, so it plays a crucial role in understanding the customer's perception of quality and influences the design of service delivery.

(2) People are a key element of the service delivery process, so it is important to select the right people. Strong employer branding can help attract the right talent, but the impact of marketing must integrate with human resources. Just as traditional marketing can help turn potential clients into clients, recruiting marketing can help turn candidates into candidates by ensuring that messages, channels, and tactics are used correctly [14].

    To manage successful delivery in an undivided environment, suppliers must find elements that can be standardized throughout the process. A great place to start is with a service plan that describes the details of the service delivery process. Although each customer has their own experience, you can make sure that the same quality standards apply to each service.

(3) to overcome the problem of variability, service providers must find the optimal balance of standardization and personalization. One way to achieve this is to

offer standardized services. For example, Xero offers its accounting software as a service in three standard packages: Starter, Standard, and Premium. By standardizing its offerings, Xero limits the variability of services by designating specific results and quality levels. At this point, variability can become a point of differentiation, providing customizable software and personalized support. Research shows that people are more likely to buy based on positive reviews.

(4) Perishable products present two main problems: distribution and pricing. For many service providers, demand can be managed through appropriate planning, such as restaurant reservations. On the other hand, growing supply can be managed through outsourcing, seasonal hiring, or adding additional locations.

When it comes to pricing, demand-based pricing structures can be implemented for services that require significant time. This is often seen in the aviation industry, as the price of a flight varies depending on demand. Suppliers can also use retention agreements to better balance supply and demand issues. Pre-contract agreements, usually used for long-term Advisory services, allow you to provide more consistent services, as well as create a predictable cash flow.

Importance of marketing services

Given the impalpability of services, their marketing becomes a particularly complex and at the same time extremely important task [15].

- Key difference: due to the growing homogeneity of the products offered, related services are becoming a key differentiator in the minds of consumers. For example: in the case of two fast food chains serving the same product (Pizza Hut and Domino's), more than the product, the quality of service distinguishes the two brands from each other. Therefore, marketers can use the service offering to differentiate themselves from competitors and attract consumers.

  Important the importance of relationships. Relationships are a key factor when it comes to marketing services. Since the product is intangible, much of the buyer's decision to buy will depend on the extent to which they trust the seller. Therefore, it is necessary to listen to the needs of the client and meet them through the appropriate offer of services and build long-term relationships that would lead to repeat sales and positive word of mouth [7].
- Customer retention. Given today's highly competitive scenario, where multiple providers compete for a limited pool of customers, retaining customers is even more important than attracting new ones. Since services are usually generated and consumed simultaneously, they actually involve the customer in the service delivery process, taking into account their requirements and feedback. In this way, they offer more customization options to meet customer requirements, which provides a higher degree of satisfaction, leading to increased customer retention [16, 17] (Table 2).

Table 2 PEST-analysis

Political factors	Economic forces
– New employment laws – Political instability in a foreign partner country – Trade control – Tax policy – The travel ban – State regulation and deregulation – Regulation of competition	– Economic growth – Changes in interest rates – Exchange rate – Inflation rate – Consumer disposable income – Unemployment rate – Fluctuation of prices – Income of potential customers in your region – Dynamics of employees ' salaries in the service sector
Social factors	*Technological factor*
– Changes in educational requirements and career settings – Population growth statistics – Age distribution – Income distribution – Attitude to customer service – Attitude to foreigners – Focus on security – Health consciousness – Lifestyle and cultural barriers – Public health and safety	– Automation – Speed of innovation – Changes in technological incentives – Computer speed/power calculation – Wireless charging – Level of innovation – Technological awareness – Internet infrastructure – Communication infrastructure – The life cycle of technology – Customer reviews on official websites, blogs, and social networks

5 Conclusion

The service sector continues to grow in size and importance for any developed country. With growth comes an endless interest in knowledge about the consumer perspective of things that underlies the study of services. The thirst for customer information seems endless at the moment. Companies like Facebook, Skype, and Google know almost every aspect of what's happening in their customers' lives. The reason is that companies want to draw conclusions from customer information to predict future behavior. Amazon, for example, predicts where a customer will make a purchase decision and ship products ahead of time to ensure fast delivery and a good customer experience. IoT also allows you to localize and interact with any product, and it will generate even more information about the user [18, 19].

Organizations will communicate with customers and employees in any physical environment, not just on the Internet. Communication is likely to take place with transmitters to communicate with various devices in order to inform customers of offers in the immediate vicinity. We have already started to see interactive ads that read faces and try to understand what is causing customers' interest. The first self-checkout history is already open, which means that every customer interaction with any product will be monitored. All these events are a bit frightening: so much information will be readily available, and we need to consider the consequences. There's not much we can do to prevent this from happening; it's like trying to prevent rain. What we can do is develop awareness and take advantage of this evolution.

Services are inherently intangible, inseparable, volatile, and perishable. Each of these characteristics creates problems for marketers and requires specific strategies. Marketers must make the intangible tangible; increase the productivity (and, in turn, profitability) of service providers; limit variability in the quality of services; and align the supply of services with changes in demand. Adapting the strategy to the unique characteristics of services, marketers can more effectively promote their offers.

References

1. Keh, H.T., Chu, S., Xu, J.: Efficiency, effectiveness and productivity of marketing in services. Eur. J. Oper. Res. **170**(1), 265–276 (2006)
2. Ostrom, A.L., Bitner, M.J., Brown, S.W., Burkhard, K.A., Goul, M., Smith- Daniels, V., Rabinovich, E.: Moving forward and making a difference: research priorities for the science of service. J. Serv. Res. **13**(1), 4–36 (2010)
3. Benoit, S., Baker, T.L., Bolton, R.N., Gruber, T., Kandampully, J.: A triadic framework for collaborative consumption (CC): motives, activities and resources & capabilities of actors. J. Bus. Res. **79**, 219–227 (2017)
4. Shostack, L.G.: Planning the service encounter. In: J.A. Czepiel, M.R. Solomon, C.F. Surprenant (Eds.), The Service Encounter. Lexington Books, New York, pp. 243–254 (1985)
5. Voorhees, C.M., Walkowiak, T., Fombelle, P.W., Gregoire, Y., Bone, S., Gustafsson, A., Sousa, R.: Service encounters, experiences and the customer journey: defining the field and a call to expand our lens. J. Bus. Res. **79**, 269–280 (2017)
6. Prasuraman, A., Zeithaml, V.A., Berry, L.L.: A conceptual model of service quality and its implications for future research. J. Mark. **49**(4), 41–50 (1985)
7. Ziyadin, S., Doszhan, R., Borodin, A., Omarova, A., Ilyas, A.: The role of social media marketing in consumer behaviour. E3S Web Conf. **135**, 04022. EDP Sciences (2019)
8. Zeithaml, V.A., Parasuraman, A., Berry, L.L.: Problems and strategies in services marketing. J. Mark. **49**(2), 33–46 (1985)
9. Vargo, S.L., Lusch, R.F.: Evolving to a new dominant logic for marketing. In: The Service-Dominant Logic of Marketing, pp. 21–46. Routledge (2014)
10. Johnson, M.D., Anderson, E.W., Fornell, C.: Rational and adaptive performance expectations in a customer satisfaction framework. J. Consum. Res. **21**(4), 665–707 (1995)
11. Leeflang, P.S., Wittink, D.R.: Building models for marketing decisions: Past, present and future. Int. J. Res. Mark. **17**(2–3), 105–126 (2000)
12. Fornell, C., Mithas, S., Morgeson, F.V., III., Krishnan, M.S.: Customer satisfaction and stock prices: high returns, low risk. J. Mark. **70**(1), 3–14 (2006)
13. Bateson, J.E.G.: Managing Services Marketing, 2nd edn. Dryden Press, Texas (1995)
14. Tukker, A., Tischner, U.: Product-services as a research field: past, present and future. Reflections from a decade of research. J. Clean. Prod. **14**(17), 1552–1556 (2006)
15. Ziyadin, S., Serikbek, A.: Internet-marketing in the sphere of higher education. Lect. Notes Netw. Syst. **84**, 290–297 (2020). https://doi.org/10.1007/978-3-030-27015-5_35
16. Belokrylova, O.S., Belokrylov, K.A., Streltsova, E.D., Tsygankov, S.S., Tsygankova, E.M.: Quality evaluation of public procurement: fuzzy logic methodology. Lecture Notes in Networks and Systems, pp. 823–833. Springer, Cham (2020)
17. Bertayeva, K., Panaedova, G., Natocheeva, N., Belyanchikova, T.: Industry 4.0 in the mining industry: global trends and innovative development. E3S Web Conf. vol. 135, p. 04026 (2019)

18. Gryzunova, N.V., Pyatanova, V.I., Ordov, K.V., Manuylenko, V.V.: Models of credit limit-setting for companies as a means of encouraging competitiveness. Entrepr. Sustain. **7**(1), 615–625 (2019). https://doi.org/10.9770/jesi.2019.7.1(43)
19. Akhmetshin, E.M., Vasilev, V.L., Murtazina, D.A., Aleynikova, O.S., Averianova, T.A., Aleksiuk, I.O.: Problems of digitalization of higher education in a small town. In: Proceedings of the 32nd International Business Information Management Association Conference, IBIMA 2018 - Vision 2020: Sustainable Economic Development and Application of Innovation Management from Regional Expansion to Global Growth, pp. 1710–1716 (2019)

Digital Modernization of the System of Public Administration and Law: Prerogatives and Practical Application

Digital Transformation as a Driver of Improvement Government Efficiency

Gulim Ismail, Sayabek Ziyadin, Aziza Zhuparova, and Raigul Doszhan

Abstract This article is dedicated to research of theoretical and methodological approaches to the problems of quality and accessibility of public services, the effectiveness of their delivery, digital transformation, as well as the relationship of digital transformation and the effectiveness of public administration. The purpose of the research is to develop proposals for the modernization of public administration in the context of digital transformation. To solve the stated research problems, general scientific methods of cognition were applied using structurally functional system and comparative approaches, content analysis of information sources, methods of factor and statistical analysis of actual and digital material. Evaluation of the effectiveness of the functioning of the public administration system in the Republic of Kazakhstan was carried out in three blocks. According to the results of evaluating the effectiveness of public administration, some problems were identified. The authors conclude that increasing the level of digitalization of public administration is closely interconnected with increasing the effectiveness of public administration, reducing corruption, and improving the conditions for doing business.

Keywords Digital transformation · Digital public administration · Quality of public services · Public administration efficiency · Openness · Customer focus · Proactivity

1 Introduction

The last decades in the Republic of Kazakhstan have been marked by numerous reforms and revisions to the principles of organizing the activities of public authorities, more and more attention is being paid to its interaction with the public, there is a growing understanding of the need to increase its degree of openness, customer focus and proactivity. Ensuring information transparency of public authorities is

G. Ismail (✉) · S. Ziyadin · A. Zhuparova · R. Doszhan
Al-Farabi Kazakh National University, al-Farabi Ave. 71, 050040 Almaty, Republic of Kazakhstan
e-mail: gulimzharaskyzy5@gmail.com

G. Mutanov and A. Serikbekuly (eds.), *Digital Transformation in Sustainable Value Chains and Innovative Infrastructures*, Studies in Systems, Decision and Control 443,
https://doi.org/10.1007/978-3-031-07067-9_15

an integral element of the concept of "quality of public administration", which includes increasing the transparency of the state apparatus and reducing the burden of government regulation for citizens.

So, in the Strategic Plan for the Development of the Republic of Kazakhstan until 2025 [1], it is noted that in order to increase the efficiency and productivity of public authorities, it is necessary to introduce digital technologies into the daily work of public authorities. At the same time, digitalization should be accompanied by information security. In parallel, work should be carried out to optimize the processes of public authorities to eliminate duplication and simplify them. The dialogue between the civil service and the public, including socio-political associations, the media, non-governmental organizations, involves the systematic informing of public activities of citizens and business entities by public authorities, as well as the development of contacts with the population and the adoption of mutually acceptable decisions on their basis. It follows that the digital transformation of public administration acts as a priority direction of state policy, which helps to minimize corruption, establish dialogue between the state, society and business, increase the level of trust between them, and create an effective system of control of citizens over the activities of public authorities.

The relevance of the study of the digital transformation of state administration in Kazakhstan is due to the presence of a large number of unresolved problems in this area. This is the need to ensure transparency of the public administration process and the effective interaction of authorities at all levels with the institutions of an emerging civil society, the realization of citizens' rights to access socially significant information, the need to increase the level of citizens' trust in public authorities, solve anti-corruption problems, and, in general, increase the effectiveness of public services in order to improve public administration. The problem of creating and increasing the transparent environment of public authorities is compounded by the inconsistency of existing national legislation with international standards, the lack of effective systems for monitoring the information openness of public authorities and their interaction with civil society. The current system does not encourage innovation, as a result of which successful changes in the work of government bodies are targeted and not widespread. Control mechanisms prevail that stimulate orientation on the process, not on the result.

Thus, we can conclude that the modernization of the public administration system in the Republic of Kazakhstan requires the search for new conceptual approaches and methods that take into account global trends and innovations in public administration.

The purpose of the research is to develop proposals for the modernization of public administration in the context of digital transformation. The goal has predetermined the resolution of the following general and particular, theoretical and applied problems:

- to research of theoretical and methodological approaches to the problems of quality and accessibility of public services, the effectiveness of their delivery, digital transformation, as well as the relationship of digital transformation and the effectiveness of public administration;

- to assessment of the public administration functioning effectiveness in the Republic of Kazakhstan;
- to identification of problems of the current public administration system in the Republic of Kazakhstan and their impact on the effectiveness of public authorities;
- to development of an effective model of interaction with central state and local executive authorities through a joint analysis of the reasons for the low quality of public services, their identification for subsequent standardization and regulation, development of recommendations for further improvement of the procedure for their provision, automation and rendering on the basis of the "one statement";
- to making recommendations on the modernization of public administration in the context of digital transformation.

To solve the stated research problems, general scientific methods of cognition were applied using structurally functional system and comparative approaches, content analysis of information sources, methods of factor and statistical analysis of actual and digital material.

2 Literature Review

Issues of the formation and reform of the public administration system, as well as the effectiveness of managerial activity of public authorities, have been the subject of numerous scientific studies. For example, the purpose of Palttala and Vos [2] research paper is to develop a framework for performance indicators to measure and improve crisis communication preparedness of public authorities. In the research of Lo Storto [3] investigating to what extent the municipality efficiency is also associated to the effectiveness of public expenditure, and consequently to quality of services offered to citizens.

Ocampo et al. [4] proposes an integrated SERVQUAL model, analytic hierarchy process, and technique for order performance by similarity to ideal solution (AHP-TOPSIS) method to evaluate service quality among employment-related government agencies. In the research of Cuadrado-Ballesteros et al. [5] seeks empirical evidence, using panel data models, of the effects of the functional decentralisation and externalisation of public services on the quality of life of the inhabitants of Spanish municipalities.

One important and essential factor that influence service quality is training. Research of Zumrah [6] is conducted to further study this relationship by exploring the possibility that transfer of training is a mediator in the relationship between training and employee service quality. Titu and Vlad [7] proposes to show evidence in measures of quality through quantification of the level of current performance of an organization, like local public administration, in accordance with performance standards.

There are few ways of efficiency quantification, but considering the comprehensive assessment need of the research problems, according to Buleca and Mura [8] studies, the most appropriate tool for performance quantification of public administration seems the packaging data analysis—DEA (Data Envelopment Analysis). In the research paper of Hwang and Akdede [9] examines whether governance quality affects public sector efficiency in the policy areas of administration, education, infrastructure, and stability. According to Matei and Enescu [10] the performance of local public administration it is an extremely actual topic and its research has a strategical importance. The main aim of this study of Cuadrado-Ballesteros et al. [11] is to analyse the effect of functional decentralisation and externalisation processes on the efficiency of local public services delivery, in order to clarify the situation. For this, 129 Spanish municipalities with populations over 10,000 between 1999 and 2007 have been considered. The authors conclude that both pure modes of public services delivery—functional decentralisation and externalisation—impact negatively on efficiency of local governments.

Among foreign scientists, such outstanding researchers as Afonso et al. [12], Asatryan et al. [13], Fakhrutdinova et al. [14], Grandy [15], Bryson et al. [16], Andrews and Van de Walle [17], Voorn et al. [18], Antony et al. [19] made a great contribution to the development of the theory of the effectiveness and quality of public administration.

To date, the concept of "digital transformation" has not been uniquely defined [20]. In a broad sense, digital transformation refers to changes in all aspects of society associated with the use of digital technologies [21]. Digital transformation is seen as a key trend that is characteristic of various industries and sectors of the economy and social sphere [22], allowing to radically increase the effectiveness or expand the volume of operations of the organization. In foreign and domestic practice and literature, the potential for the development of digital technologies is not sufficiently used and studied for management purposes (identifying problems and planning, monitoring and evaluating results, adjusting decisions). In this regard, it may be relevant and useful to assess the possibilities of using digital technologies to ensure the further development of the principle of results-based management by the state.

Thus, it can be noted that the official agenda of digitalization of public administration is still in the process of formation and is likely to require further elaboration.

3 Methodology

Evaluation of the effectiveness of the functioning of the public administration system in the Republic of Kazakhstan was carried out in three blocks: achievement of goals, interaction with citizens and organizational development of public authorities [23].

In the first block, the achievement of annual performance indicators established in strategic plans and programs for the development of territories, as well as the effectiveness of the use of the budget in achieving the goals, is assessed [24]. Assessment of the effectiveness of central government authorities is carried out according to the following criteria:

(1) achieving the goals of the strategic plan;
(2) the effectiveness of budget programs in achieving the goals of the strategic plan;
(3) the relationship of the goals of the strategic plan with budget programs.

The formula for calculating the achievement of the goals of the strategic plan and indicators of budget programs in achieving the goal:

$$R_i = (A_i + B_i) \div 2 \times C_i \tag{1}$$

where:

A_i—the coefficient of achievement of the goals of the strategic plan;

B_i—the coefficient of effectiveness of budget programs in achieving the goals of the strategic plan;

C_i—the ratio of the relationship between the goals of the strategic plan and budget programs [24].

The overall performance assessment for the "Achievement of the Goal" of the central state authorities is determined by the following formula:

$$R_{CSA} = \left[\left(\frac{R_{i1} + R_{i2} + R_{in}}{n} \right) \times 100 \right] - W \tag{2}$$

where:

R_{CSA}—final assessment of the effectiveness of achieving the goals and indicators of budget programs of the central state authorities;

R_i—the coefficient of achievement of the goals of the strategic plan and indicators of budget programs in achieving the goal;

n—number of goals of the strategic plan;

W—penalty points.

The final assessment of local executive authorities on "Achieving the goals of territorial development programs" is carried out according to the following formula:

$$R_{LEA} = (0,5 \times D) + (L + Z + G) - W \tag{3}$$

where:

R_{LEA}—final assessment of the effectiveness of achieving the goals and indicators of budget programs of the local executive authorities;

D—achievement of the objectives of the territory development program;

L—absence of violations of budgetary and other legislation following the results of inspections of budgetary development programs by state audit and financial control authorities for the evaluated period;

Z—achievement of direct results of the budget development program;
G—the effectiveness of the implementation of the budget development program;
W—penalty points.

In accordance with the result of the assessment, the degree of effectiveness of the central state authorities and local executive authorities in achieving the goals is determined.

A high degree of efficiency of central state authorities and local executive authorities corresponds to a rating indicator from 90 to 100 points, an average degree from 70 to 89.99 points, a low degree from 50 to 69.99 points. The activity of central state authorities and local executive authorities that scored less than 50 points based on the results of the assessment is recognized as ineffective [24].

In the second block, the interaction of government agencies with the population is analyzed in terms of the provision of public services, the consideration of complaints and applications, and ensuring transparency through the Open Government portals [25]. A general assessment of the effectiveness of interaction between a public authority and individuals and legal entities is determined by adding up the results of calculations in the areas of assessment:

(1) for central state authorities:

$$O = 0.5 \times U + 0.2 \times D + 0.3 \times H \tag{4}$$

(2) for local executive authorities:

$$A = 0.5 \times T + 0.2 \times F + 0.3 \times Y \tag{5}$$

where:

O—general assessment of the central state authorities under the block "Interaction of the public authority with individuals and legal entities";

A—general assessment of the local executive authorities under the block "Interaction of the public authority with individuals and legal entities";

U—assessment of the central state authorities in the direction "Quality of public services";

T—assessment of the local executive authorities in the direction "Quality of public services";

D—assessment of the central state authorities in the direction of "Openness of the public authority";

F—assessment of the local executive authorities in the direction of "Openness of the public authority";

H—assessment of the central state authorities in the direction "Quality of complaints and applications";

Y—assessment of the local executive authorities in the direction "Quality of complaints and applications".

In accordance with the result of the assessment, the degree of effectiveness of the public authority is determined. A high degree of efficiency of a public authority corresponds to a rating indicator from 90 to 100 points, an average degree from 70 to 89.99 points, a low degree from 50 to 69.99 points. The activity of the public authority that scored less than 50 points based on the results of the assessment is recognized as ineffective [25].

The third block is aimed at assessing internal processes—it analyzes the effectiveness of personnel management and the use of information technologies in the activities of a public authority [26].

A general assessment of the effectiveness of a public authority in the block "Organizational development of a public authority" is determined by the following formula:

$$Z = (0,5 \times H + 0,5 \times I) \tag{6}$$

where:

Z—general assessment for the block "Organizational development of a public authority";

H—final assessment in the direction of "Personnel Management", taking into account the deduction of penalty points in this area;

I—final assessment in the direction "Application of information technologies", taking into account the deduction of penalty points in this area.

In accordance with the result of the assessment, the degree of effectiveness of the public authorities' activity is determined according to the block "Organizational development of a public authority". A high degree of efficiency of a public authority corresponds to a rating indicator from 90 to 100 points, an average degree from 70 to 89.99 points, a low degree from 50 to 69.99 points. The activity of the public authorities that scored less than 50 points based on the results of the assessment is recognized as ineffective [26].

4 Results

The results of the evaluation are presented in Tables 1 and 2 in three blocks, as well as in the context of the evaluated public authorities.

According to the results of evaluating the effectiveness of public administration, the following problems were identified:

- the need to reformat key state planning procedures and implement the Evidence-based Public Policy tools;
- inefficiency of result-oriented budget mechanisms;
- lack of an integrated approach to budget transparency and interaction with the population;
- low level of development of budget programs;

Table 1 The results of evaluating the effectiveness of public administration in central state authorities

Central state authorities	Achievements of goals (R_{CSA})	Interaction with citizens (O)	Organizational development (Z)
ACSA	93, 47	91, 64	88, 55
MH	90, 20	48, 66	55, 97
MFA	97, 80	74, 15	39, 48
MIID	78, 13	62, 13	61, 98
MIC	89, 20	83, 17	71, 78
MSD	72, 60	74, 11	52, 64
MCS	84, 15	82, 09	56, 82
MNE	76, 53	64, 84	80, 05
MES	84, 50	56, 97	74, 76
MA	84, 97	59, 43	65, 09
MLSPP	93, 05	79, 40	89, 85
MF	91, 15	65, 92	77, 94
MDDIAI	97, 75	76, 65	83, 96
ME	83, 89	73, 86	85, 51
MJ	74, 05	78, 16	85, 74
NB	89, 40	68, 01	
GP		90, 51	
DPCA SC		77, 28	
MIA		70, 59	

(continued)

- the formal nature of regional territorial development programs;
- inconsistency of measures to digitalize public services;
- a number of problematic issues are noted regarding the organizational structure, information security, the rationality of administrative costs, and automation of internal processes;
- demand by public authorities of intermediate information and information;
- inefficiency of the current format of the Open Government reform;
- low quality and formality of government responses to appeals;
- low share of local content in the procurement of information and communication technologies.

The solution to these problems, in our opinion, is associated with the expansion of the capabilities and practice of public administration based on the results in the development and implementation of a digitalization policy for public administration, including its digital transformation [28, 29]. Studying these materials, we will see additional problems such as the formation of a base of Bank deposits as an active method of monitoring the policy of Bank deposits, assessment of the quality of public

Table 1 (continued)

Central state authorities	Achievements of goals (R_{CSA})	Interaction with citizens (O)	Organizational development (Z)
MD		48, 01	

Note
1Calculated by the author based on the formula (1), (3), (5)
2ACSA—Agency of the Republic of Kazakhstan for Civil Service Affairs
MH—Ministry of Healthcare of the Republic of Kazakhstan
MFA—Ministry of Foreign Affairs Republic of Kazakhstan
MIID—Ministry of Industry and Infrastructural Development of the Republic of Kazakhstan
MIC—Ministry of Information and Communications of the Republic of Kazakhstan
MISD—Ministry of Social Development of the Republic of Kazakhstan
MCS—Ministry of Culture and Sport of the Republic of Kazakhstan
MNE—Ministry of National Economy of the Republic of Kazakhstan
MES—Ministry of Education and Science of the Republic of Kazakhstan
MA—Ministry of Agriculture of the Republic of Kazakhstan
MLSPP—Ministry of Labor and Social Protection of Population of the Republic of Kazakhstan
MF—Ministry of Finance of the Republic of Kazakhstan
MDDIAI—Ministry of Digital Development, Innovations and Aerospace Industry of the Republic of Kazakhstan
ME—Ministry of Energy of the Republic of Kazakhstan
MJ—Ministry of Justice of the Republic of Kazakhstan
NB—National Bank of the Republic of Kazakhstan
GP—General Prosecutor of the Republic of Kazakhstan
DPCA SC—Department for the provision of courts' activity under the Supreme Court of Republic of Kazakhstan
MIA—Ministry of Internal Affairs of the Republic of Kazakhstan
MD—Ministry of Defense of the Republic of Kazakhstan

procurement: the methodology for setting credit limits for enterprises as a means of stimulating competitiveness, the mechanism for financing the agro-industrial complex also affects the digitalization of public administration [30–33].

5 Conclusion

The digital transformation of public administration is not just automation and optimization of individual processes in the provision of public functions, including the provision of public services, the introduction and use of various modern ICTs in the interests of ensuring the activities of public authorities. Digital transformation is designed to qualitatively change the content of public administration, including its individual procedures, stages of the managerial cycle, state functions, their composition and types, and such a change should lead to an increase in the quality of public administration: to ensure greater justification for government intervention (and reduce the role of the state in overall), increasing the effectiveness and efficiency of public authorities. Increasing the level of digitalization of public administration

Table 2 The results of evaluating the effectiveness of public administration in local executive authorities

Local executive authorities	Achievements of goals (R_{LEA})	Interaction with citizens (A)	Organizational development (Z)
Nur Sultan	85, 69	55,51	66, 32
Almaty	87, 21	57, 09	96, 19
Shymkent	83, 08	79, 61	75, 16
Akmola region	66, 10	72, 61	43, 13
Aktobe region	82, 87	53, 60	64, 83
Almaty region	74, 26	45, 75	41, 14
Atyrau region	84, 75	62, 17	85, 08
EKR	82, 80	91, 78	89, 80
Zhambyl region	77, 70	51, 81	64, 87
WKR	77, 33	45, 51	64, 84
Karaganda region	72, 10	55, 03	80, 51
Kostanay region	78, 98	60, 93	56, 83
Kyzylorda region	77, 73	76, 75	91, 62
Mangistau region	68, 64	44, 39	49, 24
Pavlodar region	79, 60	76, 69	50, 65
NKR	65, 61	63, 20	79, 85
Turkestan region	78, 12	52, 95	59, 48

Note Calculated by the author based on the formula (2), (4), (5)

is closely interconnected with increasing the effectiveness of public administration, reducing corruption, and improving the conditions for doing business.

An unprecedented expansion of the possibilities of working with a wide variety of data in real time (or close to it) allows government agencies to plan their activities in a completely new way, to monitor and evaluate their achievement, as well as the participation of their staff. In this sense, digital transformation is becoming a driver, a mechanism for improving the efficiency and quality of public administration.

Acknowledgements The data for the publication were collected with the financial support of the grant project AP08053346 "Research of sustainable development innovations from the point of view of their economic feasibility and building effective enterprise management in the Republic of Kazakhstan".

References

1. Legal information system of Regulatory Legal Acts of the Republic of Kazakhstan: Decree of the President of the Republic of Kazakhstan of February 15, 2018 No. 954 "On approval of the

Strategic Development Plan of the Republic of Kazakhstan until 2025 and the recognition of some Decrees of the President of the Republic of Kazakhstan as invalid (2018). http://www.adilet.zan.kz/rus/docs/U1800000636. Accessed 02 Dec 2019
2. Palttala, P., Vos, M.: Quality indicators for crisis communication to support emergency management by public authorities. J. Contingenc. Crisis Manage. **20**(1), 39–51 (2012)
3. Lo Storto, C.: The trade-off between cost efficiency and public service quality: a non-parametric frontier analysis of Italian major municipalities. Cities **51**, 52–63 (2016)
4. Ocampo, L., Alinsub, J., Casul, R.A., Enquig, G., Luar, M., Panuncillon, N., Ocampo, C.O.: Public service quality evaluation with SERVQUAL and AHP-TOPSIS: a case of Philippine government agencies. Socio Econ. Plann. Sci. (2017)
5. Cuadrado-Ballesteros, B., García-Sánchez, I.M., Prado-Lorenzo, J.M.: Effects of different modes of local public services delivery on quality of life in Spain. J. Clean. Prod. **37**, 68–81 (2012)
6. Zumrah, A.R.: Service quality in Malaysian public sector: the role of transfer of training. Proc. Soc. Behav. Sci. **144**, 111–117 (2014)
7. Titu, A.M., Vlad, A.I.: Quality indicators in reference to the evaluation of the quality management of services in local public administration. Proc. Econ. Fin. **16**, 131–140 (2014)
8. Buleca, J., Mura, L.: Quantification of the efficiency of public administration by data envelopment analysis. Proc. Econ. Fin. **15**, 162–168 (2014)
9. Hwang, J., Akdede, S.H.: The influence of governance on public sector efficiency: a cross-country analysis. Soc. Sci. J. **48**(4), 735–738 (2011)
10. Matei, A., Enescu, E.B.: Good local public administration and performance. An Empirical Study. Proc. Soc. Behav. Sci. **81**, 449–453 (2013)
11. Cuadrado-Ballesteros, B., García-Sánchez, I.M., Prado-Lorenzo, J.M.: Effect of modes of public services delivery on the efficiency of local governments: a two-stage approach. Utilities Policy **26**, 23–35 (2013)
12. Afonso, A., Schuknecht, L., Tanzi, V.: Public sector efficiency: an international comparison. Public Choice **123**(3–4), 321–347 (2005)
13. Asatryan, Z., Heinemann, F., Pitlik, H.: Reforming the public administration: the role of crisis and the power of bureaucracy. Eur. J. Polit. Econ. **48**, 128–143 (2017)
14. Elena, F., Oksana, Y., Larisa, B., Olga, Y.: Role of competence-based approach in an increase of efficiency of public administration. Proc. Econ. Fin. **23**, 1064–1067 (2015)
15. Grandy, C.: The "efficient" public administrator: Pareto and a well-rounded approach to public administration. Public Adm. Rev. **69**(6), 1115–1123 (2009)
16. Bryson, J.M., Crosby, B.C., Bloomberg, L.: Public value governance: moving beyond traditional public administration and the new public management. Public Adm. Rev. **74**(4), 445–456 (2014)
17. Andrews, R., Van de Walle, S.: New public management and citizens' perceptions of local service efficiency, responsiveness, equity and effectiveness. Public Manage. Rev. **15**(5), 762–783 (2013)
18. Voorn, B., van Genugten, M.L., Van Thiel, S.: The efficiency and effectiveness of municipally owned corporations: a systematic review. Local Gov. Stud. **43**(5), 820–841 (2017)
19. Antony, J., Rodgers, B., Gijo, E.V.: Can lean six sigma make uk public sector organisations more efficient and effective? Int. J. Product. Perform. Manag. **65**(7), 995–1002 (2016)
20. Matt, C., Hess, T., Benlian, A.: Digital transformation strategies. Bus. Inf. Syst. Eng. **57**(5), 339–343 (2015)
21. Reis, J., Amorim, M., Melão, N., Matos, P.: Digital transformation: a literature review and guidelines for future research. In: World Conference on Information Systems and Technologies, pp. 411–421. Springer, Cham (2018)
22. Gray, J., Rumpe, B.: Models for the digital transformation. Softw. Syst. Model. **16**(2), 307–308 (2017)
23. Legal information system of Regulatory Legal Acts of the Republic of Kazakhstan: Decree of the President of the Republic of Kazakhstan of March 19, 2010 No. 954 "On the System of the annual assessment of the effectiveness of the activities of central state and local executive

authorities of oblasts, cities of national importance, the capital" (2010). http://www.adilet.zan.kz/rus/docs/U100000954_. Accessed 02 Dec 2019
24. Legal information system of Regulatory Legal Acts of the Republic of Kazakhstan: Joint order of the Minister of National Economy of the Republic of Kazakhstan dated December 30, 2016 No. 540 and the Minister of Finance of the Republic of Kazakhstan dated December 30, 2016 No. 706 "On approval of the Methodology for assessing the effectiveness of achieving goals" (2016). http://www.adilet.zan.kz/rus/docs/V1600014781. Accessed 04 Dec 2019
25. Legal information system of Regulatory Legal Acts of the Republic of Kazakhstan: Joint order of the Chairman of the Agency of the Republic of Kazakhstan on Civil Service and Combating Corruption dated February 3, 2017 No. 29, the Minister of Information and Communications of the Republic of Kazakhstan dated February 6, 2017 No. 45 and the Chairman of the Committee on Legal Statistics and Special Accounts of the General Prosecutor's Office of the Republic of Kazakhstan dated February 7, 2017 No. 4 "On approval of the Methodology for assessing the interaction of a public authorities with individuals and legal entities" (2017). http://www.adilet.zan.kz/rus/docs/V1700014795. Accessed 05 Dec 2019
26. Legal information system of Regulatory Legal Acts of the Republic of Kazakhstan: Joint order of the Minister of Information and Communications of the Republic of Kazakhstan dated February 1, 2019 No. 43 and the Chairman of the Agency of the Republic of Kazakhstan on Civil Service and Combating Corruption dated February 1, 2019 No. 24. On approval of the Methodology for assessing the effectiveness of state authorities in the block "Organizational development of a state authorities" (2019). http://www.adilet.zan.kz/rus/docs/V1900018272. Accessed 06 Dec 2019
27. Center for evaluating the efficiency of activities of state authorities JSC "Institute for Economic Research": Evaluation results (2018). https://www.bagalau.kz/results/indicators/2/8/6. Accessed 06 Dec 2019
28. Streltsova, E.D., Yakovenko, I.V.: Support of dicision-making in interbudgetary regulation on the basis of simulation modeling. Smart Innovation, Systems and Technologies, vol. 139 (2019)
29. Natocheeva, N., Borodin, A., Rud, N., Kutsuri, G., Zholamanova, M., Namitulina, A.: Development of tools for realizing the potential of financial stability of enterprises. Entrepreneurship. Sustain. Issues. **7**(2), 1654–1665 (2019)
30. Bondarenko, T.G., Zhdanova, O.A., Bondarenko, N.E., Goryainova, L.V., Maksimova, T.P.: Improvement of the mechanism for financing the agro-industrial sector in the Russian federation. Int. J. Mech. Eng. Technol. **9**(9), 1419–1426 (2018)
31. Gryzunova, N.V., Pyatanova, V.I., Ordov, K.V., Manuylenko, V.V.: Models of credit limit-setting for companies as a means of encouraging competitiveness. Entrepreneurship. Sustain. Issues. **7**(1), 615–625 (2019). https://doi.org/10.9770/jesi.2019.7.1
32. Belokrylova, O.S., Belokrylov, K.A., Streltsova, E.D., Tsygankov, S.S., Tsygankova, E.M.: Quality evaluation of public procurement: fuzzy logic methodology. In: Lecture Notes in Networks and Systems. pp. 823–833. Springer, Cham (2020)
33. Bondarenko, T.G., Isaeva, E.A., Zhdanova, O.A., Pashkovskaya, M.V.: Model of formation of the bank deposit base as an active method of control over the bank deposit policy. J. Appl. Econ. Sci. **11**(7), 1477–1489 (2016)

Challenges and Risks of Digitalization of Public Administration

Galiya Dauliyeva and Aruzhan Yeraliyeva

Abstract The development of the digital economy in Kazakhstan and around the world is associated with various threats and challenges. The scope of their occurrence is extensive and difficult to predict. The article discusses the challenges and threats that arise in the development of digital public administration, which can conditionally be combined into two groups: the first—organizational, technical, and technological, the second—information security. Such a separation of emerging challenges and threats contributes to the definition and selection of forms and methods of dealing with each of them. The article discusses and summarizes the barriers to development, which are often the subject of challenges and threats. The first obstacle is that now there is not enough interaction between the various infrastructural elements of e-government, in connection with which there is a need to modernize state administrative structures. The lack of qualified personnel with digital knowledge, as well as the lack of a legislative framework in this area, is very important obstacles to the digitalization of public administration. The next barrier is cybersecurity. It should be emphasized that in the digital economy, the risks of material and moral damage from the misuse of information increase sharply. Managing emerging challenges and threats is an important part of managing the development of digitalization, but there are some features associated with the uncertainty of their occurrence, both in time and in direction and place.

Keywords Digitalization · Public administration · Cybersecurity · Challenges · Risks

G. Dauliyeva (✉) · A. Yeraliyeva
Al-Farabi Kazakh National University, al-Farabi Ave. 71, 050040 Almaty, Republic of Kazakhstan
e-mail: Galiya.Daulieva@kaznu.kz

G. Mutanov and A. Serikbekuly (eds.), *Digital Transformation in Sustainable Value Chains and Innovative Infrastructures*, Studies in Systems, Decision and Control 443,
https://doi.org/10.1007/978-3-031-07067-9_17

1 Introduction

Currently, the global economy is rapidly changing under the influence of digitalization, among other things, significant changes are taking place in the field of public administration. To keep up with the times and succeed in the development of the digital economy, effective government management of its development is important. At the moment, one can already observe a number of positive trends, the development of digital infrastructure, the use of digital and electronic technologies in everyday life, the development of e-commerce etc.

Kazakhstan adopted the state program "Digital Kazakhstan" in 2017, which is a key tool for the development of the digital economy in the Republic of Kazakhstan. The purpose of this program is to accelerate the development of the digital economy in the country, improve the digital infrastructure, and improve the quality of life of people through digitalization.

Digitalization will affect all sectors of human life, in this regard, the state program is also focused on the digitalization of the economy in all areas simultaneously. The changes will also affect the public services provision, which are also transferred to electronic format through the egov.kz platform. The provision of public services in a digital way leads to a change in the functions of the state, contributes to the transformation of institutional structures and forms of government, making them more transparent and open, and improves interaction between society and the state.

The success of the implementation of the Digital Kazakhstan program depends on the consistency of the strategic goals of the social model of socio-economic development, the vision of tasks, functions, the role of the state, society and business. Thus, the digitalization of public administration requires the development of new regulations, the transformation of existing state institutional forms [1].

Digitalization is not just a modern trend, nowadays, it is the level of digitalization in Kazakhstan that solves issues of national security, competitiveness. That is, this trend covers both domestic, internal and external problems and challenges. If we do not pay attention to the development digitalization in time, then our country can irrevocably lag behind in terms of development rates from other countries and stay away from scientific and technological progress. In this case, Kazakhstan will remain in the global economy, national security will be questionable, both individual domestic companies and the entire Kazakhstani economy in the world market will lose their competitiveness. As discussed by I. Smotritskaya, the emergence of new trends in society, politics, economy is accompanied by various challenges and threats, which are manifested in the aggregate of negative results and problems [2].

To effectively solve the emerging problems and risks associated with digitalization, the key is the state management of the digital economy, since only the state is able to regulate this process.

The description of negative results through risk parameters makes it possible to identify threats (economic, political, etc.) and their specific consequences caused by external and internal manifestations of digitalization challenges for Kazakhstan, as well as to develop risk management programs aimed at reducing possible losses from digitalization.

2 Materials and Methods

The theoretical and methodological basis of this article is the scientific work of foreign and domestic researchers and specialists in the field of public administration, management theory, theory of the digital economy, innovative management, as well as scientific and practical developments in the field of the digital economy, government programs and projects for the development of the digital economy.

The methodological basis of the research is the general scientific principles of the systems approach; the following methods of analysis were applied—strategic, logical, comparative, managerial and others; methods of qualitative and quantitative analysis of the basic directions and trends in the development of the digital economy were used.

3 Results

Society and business perform their specific functions and roles in the digitalization process, however the government has a key role in this process. Successful implementation of the "Digital Kazakhstan" program depends on the efficiency and competence of the state authorities.

According to Russian scientists E. Lenchuk, I. Smotritskaya, V. Filatov, The digital transformation of public administration involves the use of electronic platforms, which should ensure more fruitful and open interaction between the state and society, which requires the development of new regulations in this area [3].

Reforms of public administration in Kazakhstan did not always have a positive effect, due to the incompleteness of a number of tasks of previous reforms, it becomes more difficult to introduce digital technologies into public administration.

One of the problems of the digitalization of the government administration in Kazakhstan is the problem of ineffective distribution of state powers and the discrepancy between the assigned functions and the material, human and financial resources allocated to the government authorities. We can also notice that recently there have been a tendency towards an increase in funding for the activities of state bodies and the expansion of their powers.

Below are the unsolved and most important problems of digitalization of public administration (Table 1).

Table 1 Problems and goals of the digitalization of government administration

Problems	Goals [4]
• Inefficiency of budget expenditures for the maintenance of the government authorities and the performance of their functions;	• Reducing the costs of government agencies by optimizing the number of employees and the functions they perform
• Imbalance in the system of distribution of state powers, the organizational structure of state bodies and the interactions between them;	• To standardize and modernize public management and administrative processes by increasing the efficiency of public service delivery and performing control activities by civil servants
• Insufficient level of qualification of government employees in the digital sphere and low level of their professional competencies;	• Employee training, digital literacy Employees must learn not only to use digital technologies, but also to be able to train people to use digital government services
• Insufficient level of labor productivity in executive bodies, which is compensated by the excessive number of employees;	• Increasing the efficiency of spending budget funds on program (project) activities implemented by the authorities—by promoting and stimulating the introduction of digital technologies in sectors of the economy
• Lack of commitment, efficiency, and serving the public interest in the government authorities	• Increasing the level of confidence of citizens and business in the authorities and officials, support for their decisions (policies)—through the formation of an ecosystem of public and private platforms • Reduction of costs for the creation and administration of information resources and systems—due to the reuse of information technologies and services

4 Discussion

4.1 Challenges, Risks, and Threats of Digitalization of Public Administration

Referring to the Russian experience of digitalization of government administration, it can be noted that, first of all, it is necessary to organize the digitalization process of restructuring the system, structure and powers of executive authorities at the state and regional levels. According to A. Gornostaeva and A. Zakharov, in order for state bodies to meet the digitalization trends, it is necessary to legally link the powers of the authorities to material, personnel, financial resources and to the results of the performed work [5].

This goal can be achieved by developing a digital platform that will generate information from various systems, databases and resources. With the help of such a digital platform, it will be possible to determine the labor productivity of government workers, their real workload and the quality of performing their official duties, control the costs of performing certain procedures and tasks, control the timing and quality of administrative procedures.

The next step is to abandon the traditional management approach. It is necessary to understand that it is impossible to apply common standards and requirements to digital and information technologies. This is also noted by E. Rogatnykh in his work "The Impact of Digitalization on the Development of the Modern World Economy." With the transition of state bodies to digital platforms, such as egov.kz, it is necessary to change both the structure and hierarchy of public administration and the mechanisms of public administration.

Referring again to the experience of the Russian Federation, we can say that the spending government budget on the digitalization of public administration is carried out in a centralized manner. The Moscow Department of Information Technologies performs the functions of a state customer and the main manager of budgetary funds for the purchase of systems, equipment, software, services in the financial sector. To implement the same model in Kazakhstan, it is necessary to legislate the need to create state information systems, centralize the cost of digitalization, ensure the interconnection of existing and adopted decisions within the framework of the general digitalization of public administration, to introduce an evidence-based principle of spending budget funds for the purchase of digital solutions and IT systems. The model of the Moscow Department of Information Technologies can be used to form a single body for digitalization, which will act as the main customer and manager of budget funds for the purchase of programs, systems and equipment for the needs of the digitalization of the public sector. At the same time, the ability of regional bodies to spend budget funds and implement IT solutions should be limited by the right to formulate requirements.

The use of a centralized model for the distribution of budgetary funds and digitalization measures carries both positive aspects and risks. These risks include monopolization of the market for orders of digital solutions for government agencies by the largest IT companies; overestimation of the cost of development, operation and development of digital platforms, software and systems.

The introduction of a unified methodology for determining the procurement value and the introduction of digital platforms for the public sector will increase the efficiency of spending budget funds for digitalization. For this purpose, it is necessary to develop certain requirements for procurement procedures for goods, works, services in the digital sphere, as well as unify the procurement method and justify the need for procurement [6].

The idea of creating an electronic government in Kazakhstan appeared almost fourteen years ago. The egov.kz website says that during this time four large-scale stages of implementation and development of e-government have been passed. The first informational stage of this process was implemented effectively, all the necessary information about the activities of state bodies, the amount of duties, fines, a list of required documents, etc. appeared. The availability of information has greatly facilitated the life of citizens and has become a big step towards the digitalization of public administration.

However, the next steps interactive and transactional were less successful. Undoubtedly, the ability to receive certificates from different institutions without leaving home and without queuing allowed saving time and effort, reduced the time for collecting and submitting documents. Thanks to the transactional stage, it became possible to pay online utilities, penalties and fines, government fees and charges. Electronic public procurement has increased the transparency and openness of tenders and procurements.

Also, it is necessary to take into account that most of the population of Kazakhstan does not have sufficient digital literacy and is not able to independently use electronic government services and the digitalization of public administration for many does not facilitate the process of interaction with the state.

At the moment, the electronic government of Kazakhstan is at the fourth transformational stage of development. The main goal from now on is the maximum efficiency in the provision of services to citizens. However, if the problems that arose at the previous stages are not overcome, the further development of the digitalization of public administration is doubtful.

4.2 Lack of Qualified Personnel for Digitalization of Public Administration

At the moment, due to digitalization, the government of the Republic of Kazakhstan is the largest customer of IT services. The creation of an e-government platform accelerated the growth of the IT sector by 20%. However, there are reasons that inhibit the growth of the digital industry and these reasons are less obvious than the reasons for the growth of this area [7].

The most significant factors hindering the growth rate at the moment are shown in Fig. 1.

An important obstacle to the development of electronic government services is the insufficient number of computers in government institutions, as well as among the population, which leads to low efficiency in the use of previously developed information systems and leads to the inexpediency of developing new systems.

Relatively recently, buying a computer was out of reach for the majority of the country's population due to its high cost, and even now not every family has the opportunity to purchase a computer and access the Internet. This has become especially noticeable after the need for online learning arose in connection with the coronavirus pandemic. Many residents of remote villages faced the problem of lack of digital devices and Internet connection in order to be able to study, shop online, receive government services, when everything went online. There is also the problem of computer illiteracy of the population, a significant part of the population still does not have sufficient knowledge to perform any actions on digital devices and it will

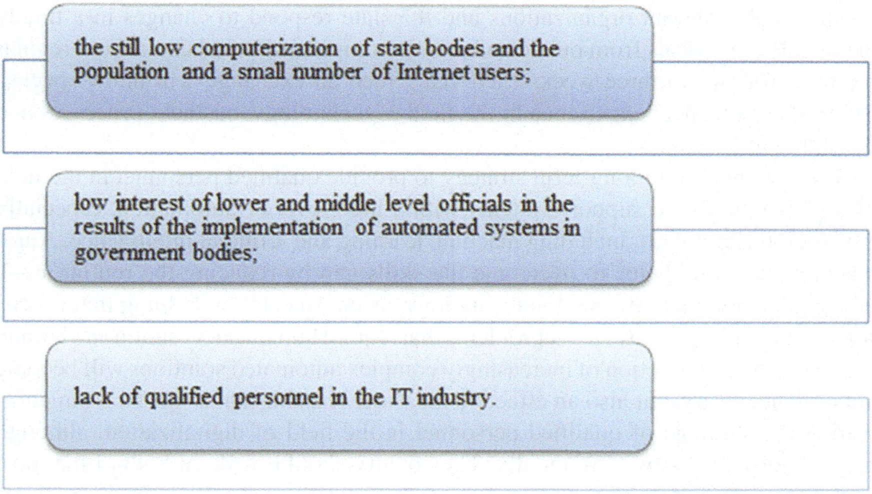

Fig. 1 Factors hindering the growth of digitalization of the public administration [7]

take several more years to improve the computer literacy of the country's population. This problem is widely covered in the state program "Electronic Government", concrete steps are outlined for its solution [8].

The more technologically advanced and knowledge-intensive the industry, the greater the impact on the success of the business is provided by staff. This statement is more true than ever for digitalization. Proper assigned work with staff affects the success of a company much more than clearly streamlined business processes. Indeed, in the end, it is with the hands and brains of employees that business processes are made and are later brought to life. Accordingly, the presence of qualified specialists is the most efficient way of increasing the likelihood of modernizing digital public administration.

The state should develop creative approaches that allow public and private companies to exchange specialists, especially when significant events arise in the field of digitalization. Digitalization is a rapidly changing field, and what is in effect today may become outdated tomorrow. It is unrealistic to think that public sector digitalisation specialists can keep up with such rapid changes in the information environment without the help of private companies. The state should develop a mechanism for digitalization professionals—especially for analysts or those who are preparing to become one—to enable them to move from state-owned companies to private companies and vice versa so that state-owned organizations can constantly update their knowledge [9].

One way to implement this idea could be to establish cooperation between government bodies with companies and other organizations, for example, universities, with the aim of recruiting digitalization specialists—operators, analysts and researchers—who will be endowed with the rights of free movement between the public and private sectors. These specialists, especially from private companies, could always be in touch to help relevant organizations and the state respond to changes in a timely manner. Professionals from public and private companies could benefit from regular rotations for two to three weeks each year. Such an exchange will allow specialists to share practical experience in the field of technology, business processes and personnel management.

The final and more long-term strategy to provide qualified personnel in the field of digitalization is to support systems with a high level of automation, especially advanced technologies, including machine learning and artificial intelligence. Automated architecture helps to overcome the skills gap by reducing the routine work of digitalization and information technology specialists and by helping them focus solely on identifying ways to fix problems that require human intervention and human analysis. The introduction of increasingly complex automated solutions will become not only necessary, but also an effective measure. Kazakhstan is still far from eliminating the shortage of qualified personnel in the field of digitalization, although the technology is improving rapidly. The country should work on both of the most important tasks at the same time: train more digitalization specialists and complicate their roles by automating routine tasks [10].

4.3 Cybersecurity

Due to the coronavirus pandemic, many services are now being delivered online, making cybersecurity issues particularly relevant. The state has developed a mobile application egov mobile, which stores digital documents of citizens, such as identity cards, driver's licenses, diplomas, health passports, and so on. Through the application, you can also get government services online, pay taxes, fines and duties, receive electronic prescriptions, find vacancies, calculate the amount of pension contributions, etc. In this regard, the government should especially ensure the cybersecurity of citizens' personal data.

Cyberattacks can be carried out with the aim of obtaining personal data of users for the purpose of further use by cybercriminals, or with the aim of disrupting an entire business process.

The government must ensure control over digital services in order to prevent cyberbullying and fraud, and it is also necessary to increase the degree of protection by allocating budgetary funds for digital security [11].

In 2017, the concept of "Cyber Shield of Kazakhstan" was adopted in Kazakhstan. The concept is aimed at maintaining the security of digital resources, electronic platforms and systems from various threats, and thereby ensure the sustainable development of our country in the context of global digitalization and competition.

In his work "Digital Technologies: Threats and Risks," V. Becher says that cybercriminals-ransomware are one of the most dangerous threats to digital security in the information environment. Such fraudsters gain access to personal computers of ordinary citizens, to secret databases of various strategic enterprises, facilities, and military bases. They can cause significant financial damage, cause irreparable harm to the country and the whole world through technological disasters. There are also cyber attacks aimed at stealing the identity of individuals, money savings from bank accounts, stealing secret documents of financial organizations and special services, stealing blockchain, assets and shares through attacks on the information bases of various companies.

The task of cybercriminals is often made easier when citizens voluntarily provide personal information, mistaking fraudsters for employees of a bank, companies, government organizations. Sometimes employees of firms, government agencies, violating job descriptions, leave documents unattended, use fake websites, lose information media, etc. The Global Risks Report [12] described the most important risks that pose global threats to humanity in the coming years. To minimize the risks and obstacles to the development of the digital economy, it is necessary to develop effective methods for assessing risks, their consequences and methods for overcoming risks. These methods should be based on IT technologies, information databases, digital platforms. In practice, the following most common groups of risk management methods have been developed and are used [13] (Fig. 2).

In order to achieve the set goals and objectives, the amount of funding and other indicators, the most suitable method is selected and there is a certain risk when choosing a method. Digital technologies do not stand still, which leads to a constant change in concepts in management practice [14]. The sooner the digital economy develops, the more risks associated with digitalization appear and these risks cannot be quantified.

Nowadays, the Internet and digital technologies have become an integral part of the life of most people, virtual reality provides tremendous opportunities and resources. With the help of the achievements of the digital economy, it is possible to optimize the procedures and processes of public administration, to increase the openness and transparency of interaction between society and the state. To minimize

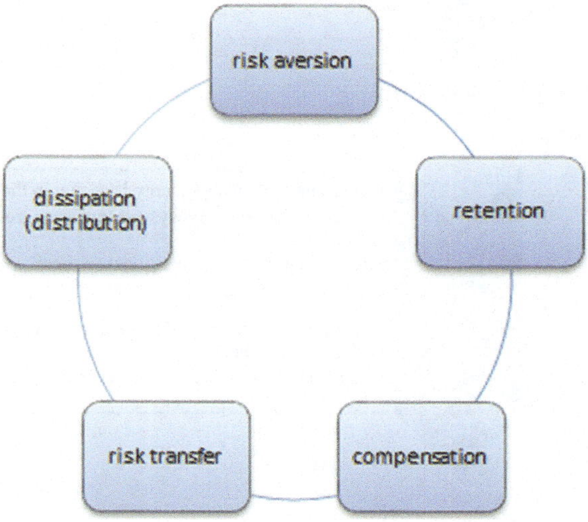

Fig. 2 Groups of risk management methods [13]

risks when using new digital technologies in public administration, it is necessary to assess and consider in advance the possible harm and appearance of digital platforms [15].

4.4 Key Areas of Cybersecurity Development

See Figs. 3, 4 and 5.

5 Conclusion

Nowadays, it is important that public administration transforms in the wake of the changing world. Five years ago, the digital revolution reached the next stage of its development—every second person on earth got access to the Internet. The challenges and problems described in this article are not only faced by the government of Kazakhstan. All over the world, governments are under pressure from all sides:

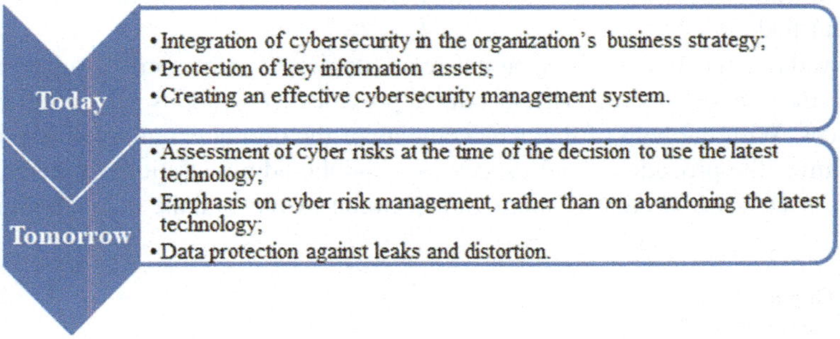

Fig. 3 Security [17]

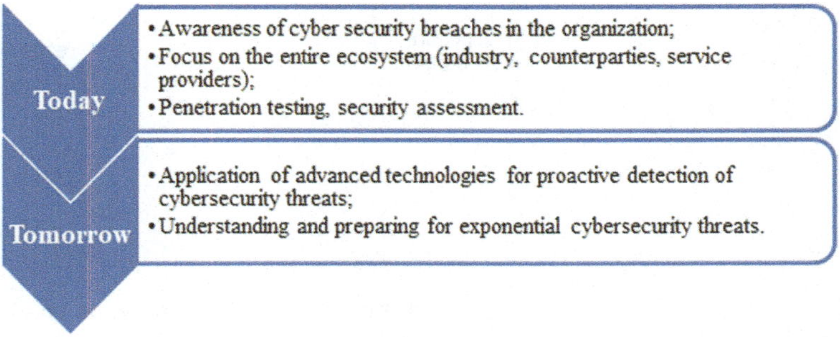

Fig. 4 Vigilance [17]

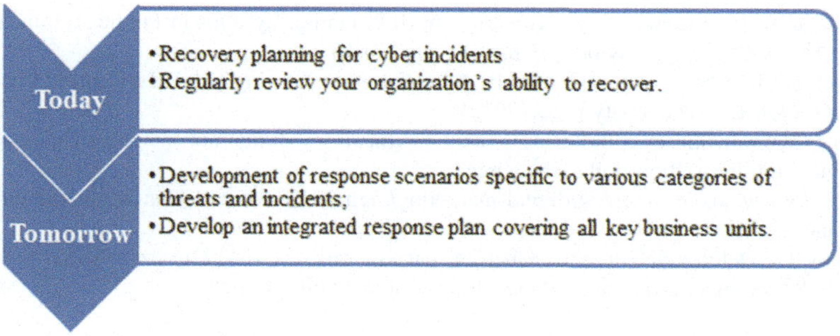

Fig. 5 Sustainability [17]

it is necessary to reduce state budget expenditures and, at the same time, improve the quality of public services provided, make more active use of digital technologies and ensure access to electronic public services for every citizen. There are both advantages and disadvantages and risks of digitalization of the public services sector. Benefits include the ability to track and resolve issues online [16], but there is a risk of cybersecurity and lack of digital literacy among a significant part of the population [18]. The government of Kazakhstan is currently facing an important task, the risks and problems indicated in this article require an effective solution.

References

1. Ziyadin, S., Suieubayeva, S., Utegenova, A.: Digital Transformation in Business, vol. 6. Scopus (2020)
2. Smotritskaya, I.: Vestnik instituta ekonomiki Rossiyskoy akademii nauk. A New Econ. Strat. Requir. New Qual. Publ. Administr. **5**, 7–22 (2017)
3. Petrov, M., Burov, V., Shklyaruk, M., Sharov, A.: CSR Report, The state as a platform. (Cyber) State Digital Econ. **3**, 35 (2018)
4. Ivanov, V., Malinetsky, G., Achieving the Sustainable Development Goals: Global Governance Challenges, vol. 1, p. 3. T.&F. Inc. (2017)
5. Akatkin, Yu., Yasinovskaya, E.: Digital Transformation of Public Administration. Datecentricity and Semantic Interoperability, pp. 8–9. DPK Press (2008)
6. Avdeeva, I., Golovina, A., Parahina, L.: Development of Digital Technologies in Economics and Management, p. 6. T.&F. Inc. (2017)
7. Halpin, E., Rankin, C., Mahtab, N.: Digital Public Administration and e-Government in Developing Nations: Policy and Practice, vol. 7. IGI Global (2013)
8. Islam, M., Ehsan, M., Global, I.G.I.: From government to e-Governance. Publ. Administr. Digital Age **4**, 12 (2012)
9. Henman, P.: Governing electronically: E-Government and the Reconfiguration of Public Administration, Policy and Power, vol. 9, pp. 45–50. Palgrave Macmillan (2010)
10. Manoharan, A., McQuiston, J.: Innovative Perspectives on Public Administration in the Digital Age, vol. 5. IGI Global (2018)
11. De Pablos-Heredero, C., Berzosa, D.: Open Innovation in Firms and Public Administrations: Technologies for Value Creation, vol. 4. IGI Global (2011)
12. U Haque, A.: Surveillance, Transparency, and Democracy: Public Administration in the Information Age, vol. 7 (2015). University of Alabama Press

13. Homburg, V.: Understanding e-Government: Information Systems in Public Administration, pp. 23–24. Routledge Taylor & Francis Group (2008)
14. Iain Gow, J.: A New Synthesis of Public Administration: Serving in the 21st Century, pp. 18–19. McGill-Queen's University Press (2012)
15. https://www.mckinsey.com/business-functions/mckinsey-digital/our-insights/public-sector-digitization-the-trillion-dollar-challenge
16. https://www.doxee.com/blog/digital-marketing/digital-transformation-in-public-administration/
17. Lipov, D.: Deloitte SNG report, Kiberbezopasnost, cifrovye riski I ugrozy (2018)
18. https://www.mgimo.ru/upload/iblock/40b/Kamolov_GSC_33.pdf

The manufacturer's authorised representative in the EU is Springer Nature Customer Service Centre GmbH, Europaplatz 3, 69115 Heidelberg, Germany. If you have any concerns regarding our products, please contact ProductSafety@springernature.com

Printed and bound by CPI Group (UK) Ltd, Croydon, CR0 4YY
28/09/2026
02224093-0002